Forschungsberichte aus dem Lehrstuhl für Regelungssysteme

Technische Universität Kaiserslautern

Band 8

Forschungsberichte aus dem Lehrstuhl für Regelungssysteme

Technische Universität Kaiserslautern

Band 8

Herausgeber:

Prof. Dr. Steven Liu

Jianfei Wang

Thermal Modeling and Management of Multi-Core Processors

Logos Verlag Berlin

λογος

Forschungsberichte aus dem Lehrstuhl für Regelungssysteme
Technische Universität Kaiserslautern

Herausgegeben von
Univ.-Prof. Dr.-Ing. Steven Liu
Lehrstuhl für Regelungssysteme
Technische Universität Kaiserslautern
Erwin-Schrödinger-Str. 12/332
D-67663 Kaiserslautern
E-Mail: sliu@eit.uni-kl.de

Bibliografische Information der Deutschen Nationalbibliothek

Die Deutsche Nationalbibliothek verzeichnet diese Publikation in der Deutschen Nationalbibliografie; detaillierte bibliografische Daten sind im Internet über http://dnb.d-nb.de abrufbar.

ISBN 978-3-8325-3699-2
ISSN 2190-7897

Logos Verlag Berlin GmbH
Comeniushof, Gubener Str. 47,
10243 Berlin
Tel.: +49 (0)30 / 42 85 10 90
Fax: +49 (0)30 / 42 85 10 92
http://www.logos-verlag.de

Thermal Modeling and Management of Multi-Core Processors

Thermische Modellierung und Thermisches Management von Mehrkernprozessoren

Vom Fachbereich Elektrotechnik und Informationstechnik
der Technischen Universität Kaiserslautern
zur Verleihung des akademischen Grades
Doktorin der Ingenieurwissenschaften (Dr.-Ing.)
genehmigte Dissertation

von
M. Sc. Jianfei Wang
geb. in Jiangsu, V.R. China

D 386

Tag der mündlichen Prüfung:	23.08.2013
Dekan des Fachbereichs:	Prof. Dr.-Ing. habil. Norbert Wehn
Vorsitzender der Prüfungskommission:	Prof. Dr.-Ing. Dr. rer. nat. habil. Wolfgang Kunz
1. Berichterstatter:	Prof. Dr.-Ing. Steven Liu
2. Berichterstatter:	Prof. Dr.-Ing. habil. Norbert Wehn

Acknowledgements

This thesis presents the results of my work at the Institute of Control Systems (LRS), Department of Electrical and Computer Engineering, at the University of Kaiserslautern.

Foremost, I would like to express my great gratitude to Prof. Dr.-Ing. Steven Liu, the head of the Institute of Control Systems, for the excellent supervision of my research, the scientific discussions, and also the good researching atmosphere. I am also thankful for offering the scholarship to me.

I would like to thank Prof. Dr.-Ing. habil. Norbert Wehn for his interest in my research and for being the second referee of my thesis. Thanks to Prof. Dr.-Ing. Dr. rer. nat. habil. Wolfgang Kunz for joining the thesis committee as the chair. Also, my deeply thanks to Prof. Dr.-Ing. Steven Liu, Prof. Dr.-Ing. habil. Norbert Wehn and Prof. Dr.-Ing. Dr. rer. nat. habil. Wolfgang Kunz for kindly supporting me with regard to my disability with dysaudia.

Thanks to the group of International School for Graduate Studies (ISGS) at the University of Kaiserslautern for offering me the scholarship and other helps.

I had a very good time working at the Institute of Control Systems. All the colleagues created an open, cooperating and warm working atmosphere, such that I can enjoy my work so much. My thanks to all them, especially to Priv. Doz. Dr.-Ing. habil. Christian Tuttas, Jun.-Prof. Dr.-Ing. Daniel Görges, Dipl.-Ing. Fabian Kennel, Dipl.-Ing. Felix Berkel, M. Sc. Hengyi Wang, Dr.-Ing. Jens Kroneis, Dr.-Ing. Liang Chen, M. Sc. Markus Bell, Dr.-Ing. Martin Pieschel, M. Sc. Michel Izak, Dipl.-Ing. Nadine Stegmann-Drüppel, Dipl.-Ing. Nelia Schneider, Dipl.-Ing. Peter Müller, Dr.-Ing. Philipp Münch, M. Sc. Sanad Al-Areqi, M. Eng. Sebastian Caba, M. Sc. Sheuli Paul, Dipl.-Ing. Stefan Simon, Dipl.-Wirtsch.-Ing. Sven Reimann, Dipl.-Ing. Tim Nagel, M. Sc. Wei Wu, Dr. Wen'an Zhang and M. Sc. Yun Wan. Thanks to the secretary Jutta Lenhardt and technicians Swen Becker and Thomas Janz. Also, thanks to my Hilfswissenschaftler M. Sc. Bicheng Chen.

My biggest thanks go to my family in China and my family in Germany for their support and love. This thesis is dedicated to them.

Kaiserslautern, January 2014

Jianfei Wang

Contents

List of Tables

List of Figures

1 Introduction

In the talk of 'There's Plenty of Room at the Bottom', the American theoretical physicist Feynman said 'Why cannot we write the entire 24 volumes of the Encyclopedia Britannica on the head of a pin?' which started the world of 'smaller' [Fey60]. From the first Intel CPU 4004 with 2000 transistors to Intel Pentium 8400EE with 2.3 billions transistors which was developed in 2010, the integration increased 100 thousand times but the size of the transistors is much smaller. However, in recent years the increase of the chip performance is slowing as transistors cannot shrink forever [Gee05]. Meanwhile, the rapid increase of information data needs higher performance processors. However, with the same power consumption, the performance of multi-core processors is much better compared with single-core processors as shown in Figure 1.1[1]. Therefore, the multi-core processor becomes the new trend of CPU development [Gee05, GK06].

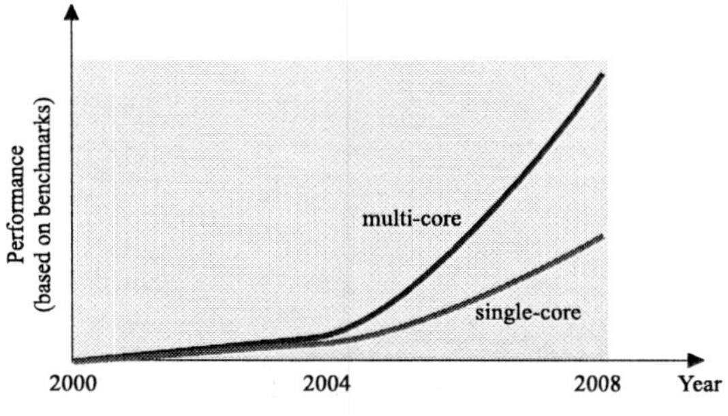

Figure 1.1: Performance improvement of multi-core processors [Gee05]

[1]The performance shown in Figure 1.1 is based on Intel tests using the SPECint2000 and SPECfp2000 benchmarks. Besides, Intel predicts that the advantages of multi-core chips will be increasing in the coming years [Gee05].

1.1 Background and motivation

1.1.1 Multi-core processors

A multi-core processor is an advanced type of processor which contains two or more cores. Each core can read and execute the program instructions independently. The multi-core processor dose not have a fixed structure, and the manufacturers design the chip differently from one another. However, the basic configuration of the multi-core processor is shown in Figure 1.2. Each core contains a processing unit and a level 1 (L1) cache. Some MCPs additionally contain a level 2 (L2) cache. All these inside components are linked to each other by an internal element interconnect bus [GK06].

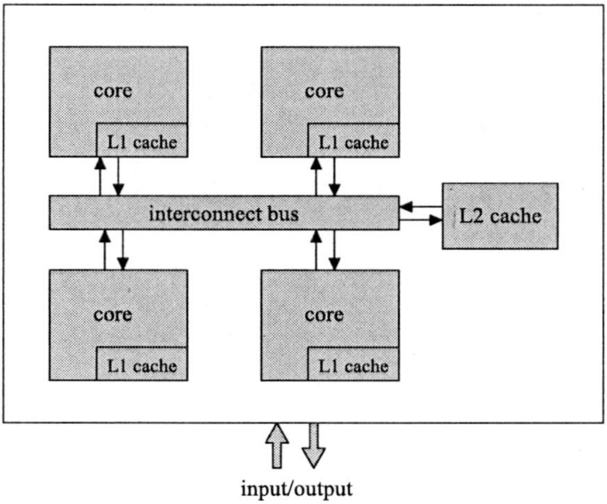

Figure 1.2: General multi-core processor structure

In comparison with single-core processors (SCP), multi-core processors have many advantages. First, the processor can handle tasks in parallel which can improve the whole process speed prodigiously. Besides, as all the cores are packed in the same chip, the communication lines are much shorter, and the communication efficiency is highly improved. Under the same power consumption, multi-core processors can have a significantly better performance than single-core processor. As shown in [Rat06], for a core, if the supply voltage is reduced by 20%, and the frequency is also reduced by 20%, the power can be reduced by about 50%. However, the core performance is only reduced by about 13%. Figure 1.3 compares the one core processor with a two cores processor. The result shows that the MCP can have much higher work efficiency with significant better performance comparing with the SCP.

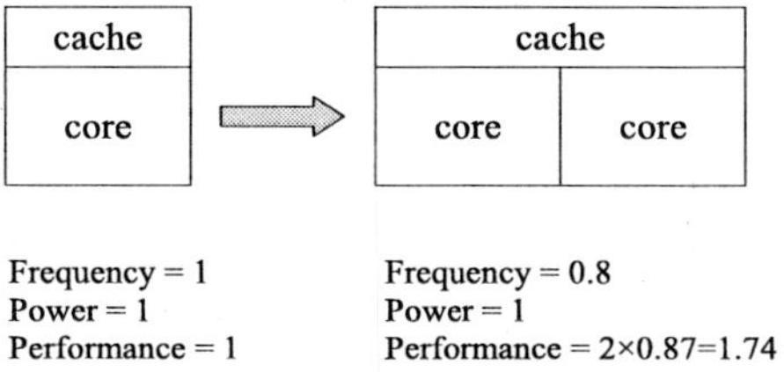

Figure 1.3: Comparison of SCP and MCP [Rat06]

From the first multi-core processor POWER4 developed by IBM in 1999 [Kah99] with two cores to Tilera's Tile-Gx8072™ processor [Til13] with 72 cores, significant progress has been made in the multi-core processor technology. Up to date, many multi-core processors are developed and are applied successfully in commercial applications, for example the IBM's CELL [PBB+05], Intel core™ i7 [Int10a, Int10b] and Tilera's Tile-Gx8072™ [Til13]. The IBM CELL processor is an 8-core MCP which was developed by IBM in collaboration with Sony and Toshiba. It consists of a dual-threaded Power Processor Element (PPE) with L2 cache and 8 Synergistic Processor Elements (SPE) with its own local cache. Intel core™ i7, which is widely used in business and high-end consumer market computers, is the name of several families of Intel desktop and laptop 64bit processors and may contain between two and six cores. Tile-Gx8072™ is a 72-core MCP which incorporates a two-dimensional array of processing elements. The cores are connected via a multiple two-dimensional mesh network.

As discussed above, multi-core processors have good performance and are widely applied. However, the power and thermal management/balancing is of increasing concerns, as it is a technological challenge to the multi-core processor development. The temperature has a significant impact on the operation reliability and chip lifespan [VWWL00, SVS06, VS06]. Thus, developing valid thermal and power management techniques for the multi-core processor becomes an urgent task currently.

1.1.2 Package technologies of chips

Nowadays, the chips are normally packed with only one layer of dies, which is called 2D package chip. However, with the development of the micro electronic techniques, the multi-core processor structure becomes more complicated. The traditional 2D package technology suffers from the fact that the complex interconnect network consumes a lot of energy and is accompanied with a significant amount of heat, which may be fatal to the chip. Therefore, in recently years multi-die stacking technology (3D stacked package) emerges due to the requirement.

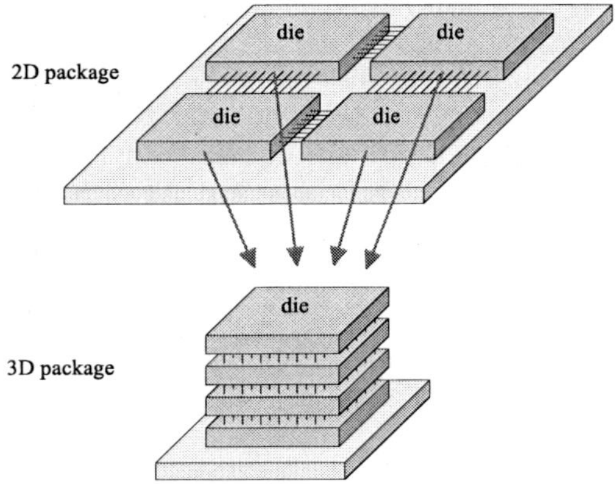

Figure 1.4: 2D package and 3D package

The 3D package is a technology which stacks at least two dies together, which means a chip contains at least 2 layers of dies as shown in Figure 1.4. Every two layers are directly connected using through-Silicon via (TSV) [Mot09]. Compared with the 2D package, this technology has many advantages [AAF98]. The size, the weight and the footprint of the chip can be highly reduced. Besides, as a result of the reduction of the interconnect length, the signal delay, the noise, and the power consumption can be decreased. Meanwhile, the chip can run at a faster rate under the same power consumption, and the interconnect bus bandwidth can be increased. However, because of the stacking, much more heat will be gathered inside the dies as the power density has a linear relationship with the number of the stacked layers. This is one of the main challenges of the power and thermal management techniques in the 3D chips.

1.1.3 Thermal model techniques

Two types of heat transfer are considered, namely heat conduction and heat convection, while the radiation heat transfer is negligible [WLC03]. The basis of heat conduction is Fourier's law [Kre00]. According to the Fourier's law, the rate of heat transfer through a material is proportional to the negative temperature gradient in any direction [Kre00]. The heat convection rate between the multi-core processor and the environment or the liquid cooling system can be described by Newton's law of cooling [AB10], i.e. the heat transfer rate of two different media is proportional to the thermal difference between the two media.

Based on the above two theories, the thermal behavior normally can be modeled in two types. A simple solution is to generate the equivalent thermal resistance-capacity

(R-C) circuit, and describe the heat transfer process as an R-C network as proposed in [SAS02] and [SSH+03a]. This model is inspired by the fact that the heat transfer and the electrical phenomena are dual [Kre00]. Figure 1.5 shows the heat conduction between two blocks of an object which is insulated and dose not contain internal heat source. As shown in the Figure, the heat transfer model between the blocks can be approached by the heat resistance R and heat capacity C. The transfer rate between the two blocks depends on the temperature difference $T_1 - T_2$ between the two blocks.

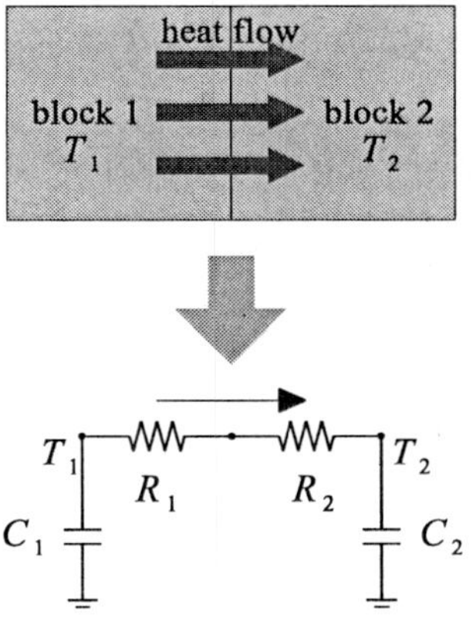

Figure 1.5: An insulated heat transfer between two blocks

Therefore, by modeling each block with thermal resistances and a capacity, and by connecting different blocks with thermal resistances, the whole thermal behavior of the multi-core processor can be approached by an R-C network. Afterwards, an ordinary differential equation (ODE) based thermal model can be achieved. A linear state space model based on the ordinary differential equation is given in [MMA+08] and [ZAD09], and the state vector is defined as the temperatures of the cores [MMA+08] or the divided blocks of the chip [ZAD09].

Another solution is to model the system by the partial differential equation (PDE). According to Fourier's heat conduction law, the dynamical heat transfer process of the multi-core processor is described as a three dimensional (3D) PDE model with boundary conditions. Therefore, a PDE based modeling technique can describe the thermal behavior more precisely. In [MSS+05], a 3D PDE with two different layers is introduced to present the analytical thermal model of the processor and then the model is transformed to an infinite ODE system via Duhamel's Theorem and Laplace transformation. In [CRT98], a 3D/1D mixed model strategy is introduced. The thermal behavior of the die is modeled as a 3D PDE while the two heat diffusion paths are treated as 1D thermal resistances in order to reduce the model complexity. With the 3D PDE model the whole die temperature distribution can be achieved.

Under the R-C network modeling method, the whole block has a uniform temperature. Hence, the temperature distribution of each point can not be obtained, and the precision is lower than under the PDE model. However, for a system with irregular volume, the PDE model is overly complex, which leads to the difficulties in the control design. For this kind of systems, the ODE R-C model is more suitable.

1.1.4 Power and thermal management methods

There are two possible ways at present to optimize the power/thermal behavior, which are employed in different stages. One is in the earlier processor hardware design stage. The thermal and power optimization can be developed with designing the lower power CMOS circuit [DMNH10] and optimizing the on chip layout [CCC12, CRAI13]. Another way is the real time online dynamic thermal/power management technique. After the dynamic voltage and frequency scaling (DVFS) technology introduced in 90's [MDVPO90], the real time power and thermal control becomes a new trend of thermal and power management technology. The DVFS technology can be implemented with the phase locked loop shown in Figure 1.6.

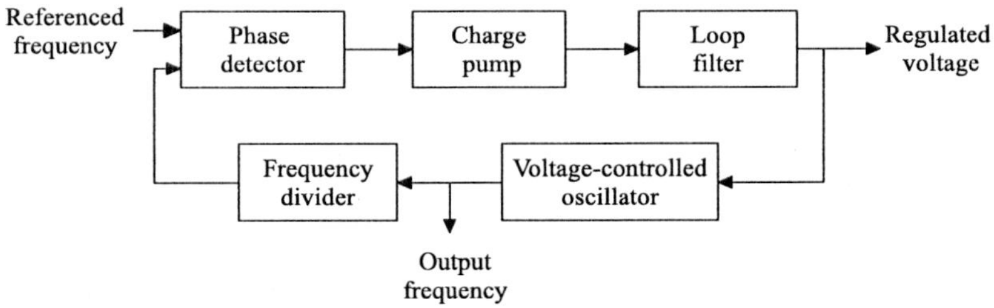

Figure 1.6: Phase locked loop

Nowadays, due to the developed techniques the DVFS can support very small voltage adjustment within extremely short time compared with the heat transfer process. For instance, the MC13783 power management chip can adjust the voltage with a step of 25 mV in some μs [Fre09]. This technology is widely applied in the multi-core processor design. The technique can be approached by two different schemes, namely distributed (per core) DVFS and global DVFS . The power consumption and thermal balancing can be controlled via appointing the optimized supply voltage and frequency online.

Some DVFS based ways for power and thermal management have been developed. The structure of the temperature and power management policy is shown in Figure 1.7. Generally, the real time task assignment and the temperature of the cores are both included in the control design.

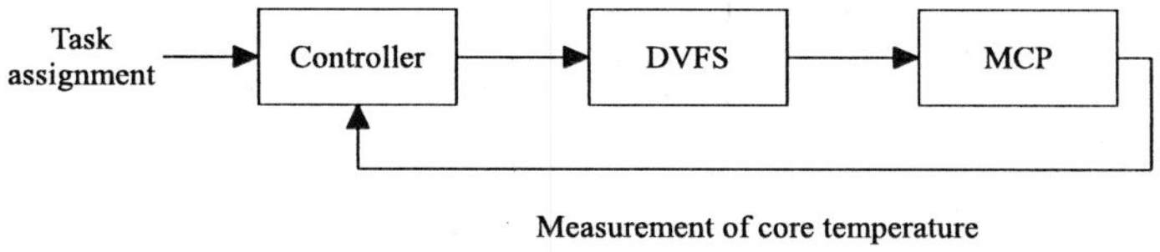

Figure 1.7: Closed loop control structure

In [WB12], a practical DVFS thermal management policy is proposed and the temperature is controlled by a logic algorithm. A proportional controller with saturation is proposed in [FKLK12] to ensure that the maximum temperature of the cores follows the thermal set-point under utilization constraints. In [LFQ12], a predictive dynamic thermal management algorithm is designed, and the control aim is to maximize the system throughput, meanwhile, guaranteeing that the temperature of the core will not exceed the constraints. In [MMA$^+$08] and [ZAD09], an ODE linear state space model is proposed and an optimal control technique is introduced to decrease the temperature gradients and hot-spots, and to keep the cores under the maximum temperature limitation. A hybrid local-global thermal management technique is proposed in [JM09a] and [JM09b]. The thermal and power management policy contains a neural network classifier to filter the thermally unsafe configurations, a high speed model to predict the system performance and an intelligent search algorithm to get the control decision.

As mentioned before, much more heat will be gathered inside the dies of 3D stacked package multi-core processor compared with the 2D package MCP. The cooling problem can be solved by a micro-channel liquid cooling system which is installed between every two layers [CAAR09, Kin12]. Because of the complexity of the forced convection heat transfer and the micro size of the channel, the modeling of thermal behavior is approached by an R-C model [CAAR09, MYL09, CAR$^+$10, ASP$^+$09]. An R-C model of the dies with a multi-pump is developed in [ZAD13]. However, in this model, the thermal behavior of the liquid cooling system is not considered. The thermal management of the 3D ICs is a new developing area. An adaptive dynamic frequency scaling technique for the 3D MCPs can be found in [CPLK12]. Some operation policies based on task scheduling and DVFS technology can be found in [CAA$^+$09, LY09, ZXD$^+$08]. In [ZAD13], an optimal thermal management policy is given by optimizing the power consumption under a thermal R-C model.

1.2 Objectives and structure of the dissertation

1.2.1 Objectives and contributions

Objectives

The objective of this dissertation is to develop feasible modeling and control techniques to manage the thermal behavior and power consumption of multi-core systems. The main tasks in this dissertation are:

1. Modeling the thermal behavior of 2D package MCP systems
2. Optimal thermal and power management policy design for 2D package MCP systems
3. Modeling the thermal behavior of 3D stacked package MCP systems with a micro-channel liquid cooling system
4. Optimal thermal and power management policy design for 3D stacked package MCP systems

Contributions

As shown in Section 1.1.3 literature proposes an R-C network as well as a 3D PDE modeling approach for modeling the thermal behavior of multi-core processors. As the 3D PDE model is not suitable for the control design, the thermal behavior of the multi-core processor is modeled by a group of 1D PDEs where each 1D PDE models the thermal behavior of one core. Previous control approaches to manage the power and temperature of the cores are based on an R-C network modeling approach [MMA+08, ZAD09]. However, this thesis proposes an optimal control policy based on the 1D PDEs model.

Further, the thesis considers the 3D stacked package MCP with a micro-channel liquid cooling system. All existing modeling approaches use the R-C network modeling approach. In this dissertation the 3D stacked package MCP including the micro-channel liquid cooling system is modeled in a similar way as in [MYL09]. However, the model of micro-channel liquid cooling system is simplified compared to [MYL09] in order to design a controller which is suitable for implementation. A novel two step controller is designed to manage the temperature and the power of the system. Besides, a robust observer is designed to estimate the system states and a stability condition for the whole system is given.

1.2.2 Structure of the dissertation

The rest of the dissertation is structured as follows.

In Chapter 2, the power dissipation of the die is modeled, the heat conduction in the die and heat convection between the chip and the environment are modeled. The Fourier's law of heat conduction and Newton's law of cooling are employed to get a 3D PDE model of the chip. The Sturm-Liouville theory is proposed to solve the 3D PDE and obtain the temperature distribution. Besides, a simulation example is presented in this chapter to test the model. Chapter 2 is the foundation of Chapter 3 for the 2D package system.

In Chapter 3, the 3D PDE model is transformed to a group of 1D PDEs. The thermal behavior of each core is described by a 1D PDE. The heat influence among the cores is obtained with the core boundary temperature gradient. A cost function is introduced which weights the difference of the temperature among the cores and the power consumption. A PDE based optimal control policy is proposed to manage the power and temperature of the cores. The system input vector is the power consumption of the cores, which can be applied to obtain the supply voltage and operation frequency assigned to each core via the DVFS technology.

The advanced 3D stacked package chip is introduced in Chapter 4. The micro-channel liquid cooling system is considered to cool the chip inside. Different from normal size channel, the micro-channel has some special features. The 3D PDE based micro-channel fluid dynamic and heat transfer model is presented in this thesis. The liquid flow and thermal characteristics are analyzed. The liquid in the channel is divided into blocks and modeled by ODEs. The heat transfer of the dies is modeled by an R-C network. A simple pump model is also given in this chapter. Further, integrating the model of the dies, the channels and the pump, the system is modeled as a nonlinear ODE system.

Based on the model developed in Chapter 4, a thermal and power management policy is proposed in Chapter 5. The control design of the system contains two steps. The liquid velocity is chosen from a set of constant velocities and a logic algorithm is developed to determine the liquid velocity. Based on this control policy, the system is modeled as a linear switched system where each subsystem refers to the model under one liquid velocity. In the second step a model predictive controller is designed to balance the temperature among the cores. To reach the control object, a cost function is introduced to weight the difference of the temperature between the cores and the control input. As not all necessary states can be assumed to be measurable and the system contains unknown inputs like the unknown parts of the power consumption of the dies, a robust H_∞-observer is employed to limit the influence of the unknown inputs to the estimated state vector. Further, a theorem is proposed to prove the stability of the controlled switched system under the separate control and observer design. Finally, the results are verified by simulation.

2 Modeling of the thermal behavior of the die

As discussed in Section 1.1.3, both an R-C network and a PDE modeling approach can be employed to describe the thermal behavior of the die. In case of the R-C network modeling approach a numerical method is applied to discretize the die into finite elements and then describe each element as an R-C element. Thus, the whole system can be described as an R-C network [Kre00]. The other method models the thermal behavior by a PDE and corresponding boundary conditions based on the Fourier's heat transfer theory and Newton's law of cooling [Geb71]. The PDE model can be solved based on the eigenvalue and eigenfunction system [MÖ94]. The PDE is projected to a group of infinite number ordinary differential equations based on the eigenfunctions to get the thermal behavior on a particular eigenvalue and eigenfunction direction. With a superposition of these signal behaviors, the thermal behavior of the whole die can be obtained.

In this chapter, the power consumption of the die is analyzed, then a PDE based analytical heat transfer model is presented, and the model is solved by the Sturm-Liouville theory. An illustrative example based on MATLAB simulation completes this chapter.

2.1 Power consumption analysis

At the circuit level a core, also called microprocessor, consists of a logic circuit which is made up by the three basic gates, namely AND, OR and NOT gates. These gates are built up by transistors, see Figure 2.1. Each transistor has two states, which is on and off, to represent the two binary value 1 and 0 respectively. For a more detailed description of the design of microprocessors see [Hwa06].

Generally, the instantaneous power consumption of a gate contains two parts, namely dynamical power consumption and leakage power consumption. The dynamical power consumption happens during the process that the transistor switches on/off. The leakage power consumption is the static power consumption, which happens when the chip power supply is on. The power consumption of a gate depends on the supply voltage $V_{\mathrm{dd}i}$ and working frequency f_i of the core i.

As shown in Figure 2.1, the dynamical power consumption consists of two parts. One part is caused by the load charging/discharging when the PMOS and/or NMOS transistor switch on/off. PMOS is a p-type MOSFET (metal-oxide-semiconductor field-effect transistor) and NMOS is an n-type MOSFET. The switching current I_{switch} leads to the switching power consumption $P_{i,\text{switch}}$. The other part happens at the moment that both the PMOS and the NMOS are on, a short current I_{ci} occurs between the supply voltage $V_{\text{dd}i}$ and the ground, causing a power consumption, which is the second part $P_{i,\text{short circuit}}$ of the dynamical power consumption. The leakage power consumption $P_{i,\text{leakage}}$ is caused by the leakage current. Therefore, the power consumption of each core is divided in three parts [CB95, VS06]

$$P_i\big(f_i(t), V_{\text{dd}i}(t)\big) = P_{i,\text{switch}} + P_{i,\text{short circuit}} + P_{i,\text{leakage}}. \tag{2.1}$$

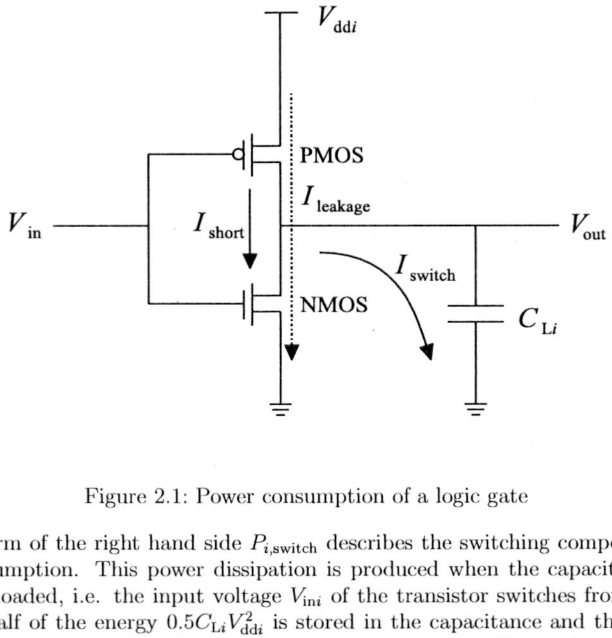

Figure 2.1: Power consumption of a logic gate

The first term of the right hand side $P_{i,\text{switch}}$ describes the switching component of the power consumption. This power dissipation is produced when the capacitance $C_{\text{L}i}$ of the gate is loaded, i.e. the input voltage $V_{\text{in}i}$ of the transistor switches from 0 to $V_{\text{dd}i}$, therefore, half of the energy $0.5C_{\text{L}i}V_{\text{dd}i}^2$ is stored in the capacitance and the other half dissipates in the logic cell as heat and the output voltage is $V_{\text{out}i} = V_{\text{dd}i}$. When the voltage $V_{\text{in}i}$ switches from $V_{\text{dd}i}$ to 0, the energy stored in the capacitance dissipates in the logic cell and $V_{\text{out}i} = 0$ [Har06]. Summing up the switching power of all gates results in the switching power of the core given by

$$P_{i,\text{switch}} = \alpha_{0\to 1,i} C_{\text{L}i} V_{\text{dd}i}^2 f_i, \tag{2.2}$$

where f_i is the operation frequency of core i, $\alpha_{0\to 1,i}$ is defined as the average number of power consuming transition in the core within a clock cycle with the capacitance $C_{\text{L}i}$ [CB95].

The second term $P_{i,\text{short circuit}}$ in (2.1) represents the short circuit component of the power. When both the PMOS and NMOS transistors are active during the input voltage switching as shown in Figure 2.2 a short current I_{short} flows from the power supply to the ground. Defining the short current I_{ci} of the whole core i as the sum of the short currents of all gates the resulting power consumption caused by the short current is given by [CB95]

$$P_{i,\text{short circuit}} = I_{ci} V_{\text{dd}i}. \tag{2.3}$$

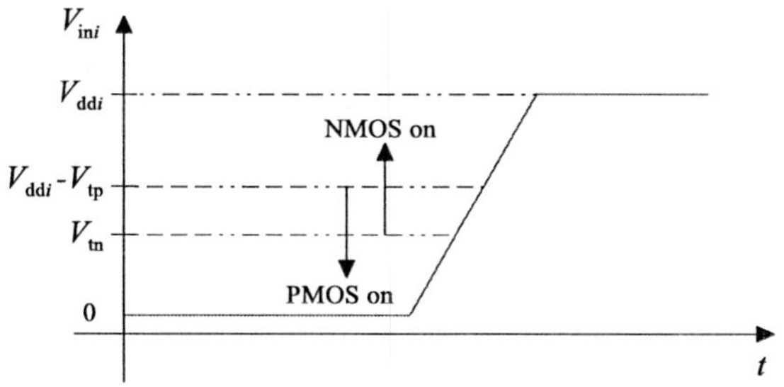

Figure 2.2: Input voltage switching

The third term in (2.1) is the leakage power consumption $P_{i,\text{leakage}}$. It contains the gate leakage, source/drain junction leakage and subthreshold leakage [AFP04]. The leakage power consumption has the form

$$P_{i,\text{leakage}} = I_{\text{l}i} V_{\text{dd}i}, \tag{2.4}$$

where $I_{\text{l}i}$ is leakage current of the whole core i, i.e. the sum of the leakage currents of all gates.

In the last two decades, dynamical voltage and frequency scaling (DVFS) technology has been developed to manage the dynamical power and temperature of the processor. According to [Bak10], the supply voltage and the maximum core frequency have the following relationship

$$f_{\text{max}i} \propto V_{\text{dd}i}, \tag{2.5}$$

where $f_{\text{max}i}$ is the maximum frequency that can be applied on the core i under the supply voltage $V_{\text{dd}i}$. Therefore, the voltage can be changed according to the reference frequency as shown in Figure 1.6.

2.2 Heat transfer analysis of multi-core processors

The multi-core processor cross-sectional view is shown in Figure 2.3. There are two heat

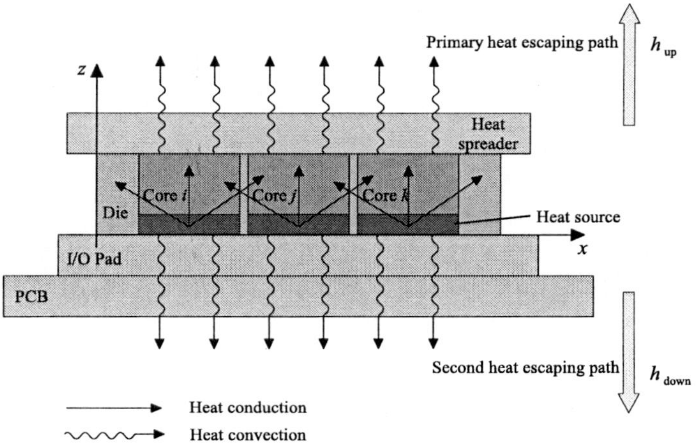

Figure 2.3: Cross-sectional view of a Multi-core processor

escaping paths [CRT98, WLC03]. The first path is from the top of the chip, which is composed of the thermal interface material, heat spreader and heat sink. The second path is from the bottom of the chip, which contains the input/output pads and the print circuit board. The heat which spreads from the four sides is ignored as the area of these sides is very small compared with the area of the top and bottom [CRT98, HL09]. The heat source is on the bottom of the die where the integrated circuit is printed.

A mixed 3D/1D strategy is introduced to model the heat transfer of the chip [CRT98] in two steps.

1. A 3D model is employed for the die to achieve a high degree of accuracy.
2. Two heat spreading paths are treated as 1D thermal resistances to reduce the computational complexity.

Figure 2.4 shows the die of the multi-core processor in Figure 2.3. According to Fourier's heat conduction law [Fou09], the heat transferred per time unit dQ/dt [J/s] through an oriented infinitesimal surface area element dS [m^2] of a material is proportional to the negative gradient of the temperature $T(x, y, z, t)$ [K] along the surface outward normal direction, i.e.

$$\frac{dQ}{dt} = -K\frac{\partial T(x, y, z, t)}{\partial \boldsymbol{n}}dS, \tag{2.6}$$

where $\partial T(x, y, z, t)/\partial \boldsymbol{n}$ [K/m] is the temperature gradient along the surface outward normal direction $\boldsymbol{n}$ and K [W/(m $\cdot$ K)] is the thermal conductivity of the die. Since the heat is always flowing from the higher temperature side to the lower temperature side,

dQ/dt and $\partial T(x,y,z,t)/\partial \boldsymbol{n}$ should be with opposite signs.

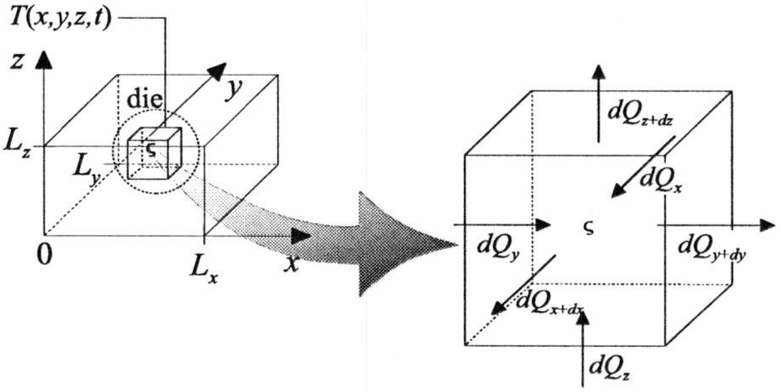

Figure 2.4: Heat conduction of the die

Selecting the total closed surface ς [m^2] as shown in Figure 2.4 [1], the whole heat variation Q [J] (caused by conduction and internal heat sources) in the closed volume in the time interval t_1 to t_2 is

$$Q = \int_{t_1}^{t_2} \iint_{\varsigma} K \frac{\partial T(x,y,z,t)}{\partial \boldsymbol{n}} dS dt + \sum_{i=1}^{N} \int_{t_1}^{t_2} \int_{x}^{x+dx} \int_{y}^{y+dy} \int_{z}^{z+dz} q_i(x,y,z,t) dx dy dz dt, \tag{2.7}$$

where $q_i(x,y,z,t)$ [W/m^3] is the internal heat generation function per time unit and per volume unit of core i and N is the number of cores in the processor. The first term of the right hand side of equation (2.7) indicates the overall heat which goes inside the closed volume (e.g. the cube in Figure 2.4) from t_1 to t_2 while the second term expresses the internal heat source in the closed volume. If $q_i(x,y,z,t)$ is in the heat source area of a running core i, $q_i(x,y,z,t)$ is nonzero, otherwise $q_i(x,y,z,t) = 0$.

Remark 2.1. *In equation* (2.7)*, only the heat source of the cores is included. For some MCPs which contain some other parts, e.g. L2 cache, the heat source of these parts also need to be considered as internal heat sources.*

To increase/reduce the temperature of the material in a closed volume from $T(x,y,z,t_1)$ to $T(x,y,z,t_2)$, the energy

$$\int_{x}^{x+dx} \int_{y}^{y+dy} \int_{z}^{z+dz} \sigma\rho \Big(T(x,y,z,t_2) - T(x,y,z,t_1) \Big) dx dy dz \tag{2.8}$$

is required, where σ [J/(kg $\cdot$ K)] is the specific heat capacity of the die and ρ [kg/m^3] is the density of the die. According to the energy conservation law, the value of (2.8)

[1] In Figure 2.4 for instance the total closed surface ς is the surface of the cube, i.e. here the surface area is six times the area of the square, the side of the cube.

should equal to Q in (2.7). Applying the Gaussian theory (see A.1) on (2.7), and setting the result equal to (2.8) results in

$$\int_{t_1}^{t_2}\int_{x}^{x+dx}\int_{y}^{y+dy}\int_{z}^{z+dz}\left[K(\frac{\partial^2 T}{\partial x^2}+\frac{\partial^2 T}{\partial y^2}+\frac{\partial^2 T}{\partial z^2})+\sum_{i=1}^{N} q_i(x,y,z,t)-\sigma\rho\frac{\partial T}{\partial t}\right]dxdydzdt=0. \tag{2.9}$$

In equation (2.9) the integral of the first summand indicates the energy variation caused by conduction and the integral of the second summand indicates the energy variation caused by heat sources. The sum of those two integrals is equivalent to the internal energy variation.

As t_1, t_2 and the closed volume with the surface ς are arbitrarily defined, we get the PDE

$$\sigma\rho\frac{\partial T}{\partial t}=K\left(\frac{\partial^2 T}{\partial x^2}+\frac{\partial^2 T}{\partial y^2}+\frac{\partial^2 T}{\partial z^2}\right)+\sum_{i=1}^{N} q_i(x,y,z,t) \tag{2.10}$$

with respect to the coordinate system defined in Figure 2.4, see also Figure 2.3.

According to Newton's law of cooling [AB10], the heat transferred per time unit between the material of a closed volume and the environment through an infinitesimal surface area element dS is proportional to the temperature difference between the closed volume and the environment which surrounds it, i.e.

$$\frac{dQ}{dt}=h_k\Big(T(x,y,z,t)-T_\infty(x,y,z,t)\Big)dS, \tag{2.11}$$

where $T_\infty(x,y,z,t)$ is the ambient temperature on the boundary and h_k $[\mathrm{W}/(\mathrm{m}^2\cdot\mathrm{K})]$ is the convective heat transfer coefficient of the kth surface area element.

In this thesis, an ashlar-formed closed volume as in Figure 2.4 is considered. The convective heat transfer coefficient at the upper surface is given by h_{up} and at the lower surface by h_{down}. As mentioned before, it is assumed that there is no heat escaping from the four sides, i.e. $h_k=0$.

Substituting (2.6) into (2.11), the general boundary condition is given by

$$-K\frac{\partial T(x,y,z,t)}{\partial \boldsymbol{n}}=h_k\Big(T(x,y,z,t)-T_\infty(x,y,z,t)\Big). \tag{2.12}$$

Therefore, the boundary conditions in x- and y-direction can be described as

$$K\frac{\partial T(0,y,z,t)}{\partial x}=0, \qquad K\frac{\partial T(L_x,y,z,t)}{\partial x}=0 \tag{2.13}$$

in x-direction where L_x is the die size in x-direction, see Figure 2.4, and

$$K\frac{\partial T(x,0,z,t)}{\partial y}=0, \qquad K\frac{\partial T(x,L_y,z,t)}{\partial y}=0 \tag{2.14}$$

in y-direction where L_y is the die size in y-direction, see Figure 2.4.

The boundary condition of the primary heat transfer path is

$$K\frac{\partial T(x,y,L_z,t)}{\partial z} = -h_{\text{up}}\Big(T(x,y,L_z,t) - T_\infty(x,y,L_z,t)\Big), \tag{2.15}$$

where L_z is the die size in z-direction, see Figure 2.4 and also Figure 2.3, and the boundary condition of the second heat spreading path is

$$K\frac{\partial T(x,y,0,t)}{\partial z} = h_{\text{down}}\Big(T(x,y,0,t) - T_\infty(x,y,0,t)\Big). \tag{2.16}$$

The right hand sides of (2.15) and (2.16) have opposite signs, as the heat convection of these sides are to opposite directions.

Thus, the thermal behavior of the die is described by the PDE (2.10) with the boundary conditions (2.13)-(2.16).

2.3 Solution of the unsteady-state heat conduction

The transient thermal behavior[2] of the die can be obtained by solving the 3D heat conduction PDE presented in the previous section by the Sturm-Liouville eigenvalue system, the separation-of-variables method and the integral transform [Hol86, Ölç64].

First a Sturm-Liouville eigenvalue system is employed based on the homogeneous function of the heat conduction equation (2.10) and the boundary conditions (2.13)-(2.16). The eigenvalue system can be solved by the separation-of-variables method. With each eigenfunction as the kernel, a three-dimensional finite integral transform of the heat distribution $T(x,y,z,t)$ is applied and it transforms the 3D PDE to an ODE. By solving the ODE and then transforming back to the 3D system, the unsteady-state heat distribution can be achieved.

The Sturm-Liouville eigenvalue system [Ölç64, MÖ94, Ch. 3] (see A.2 for details) is introduced as

$$\nabla^2\phi_{abc}(x,y,z) + \lambda_{abc}^2\phi_{abc}(x,y,z) = 0, \tag{2.17}$$

with the boundary conditions

$$K\frac{\partial\phi_{abc}(x,y,z)}{\partial \boldsymbol{n}} + h_k\phi_{abc}(x,y,z) = 0, \tag{2.18}$$

where $\nabla^2 = \partial^2/\partial x^2 + \partial^2/\partial y^2 + \partial^2/\partial z^2$ is the Laplace operator. Equation (2.18) is the summary of the six boundary conditions. The solution of the eigenvalue problem has

[2]In the research area of heat transfer the term *unsteady-state conduction* is used to describe the transient heating or cooling process before an equilibrium is established [Hol86, Ch. 4] and [Geb71, Ch. 3].

the general form

$$\phi_{abc}(x, y, z) = \phi_{xa}(x)\phi_{yb}(y)\phi_{zc}(z) \tag{2.19}$$

and

$$\lambda_{abc}^2 = \lambda_{xa}^2 + \lambda_{yb}^2 + \lambda_{zc}^2. \tag{2.20}$$

by separating the variables, see [MÖ94, Sec. 3.1]. The eigenfunction $\phi_{zc}(z)$ with respect to z is given by

$$\phi_{zc}(z) = K\cos(\lambda_{zc}z) + \frac{h_{\text{down}}}{\lambda_{zc}}\sin(\lambda_{zc}z), \tag{2.21}$$

see [MÖ94, Sec. 3.1] for the detailed derivation. For the eigenfunctions $\phi_{xa}(x)$ and $\phi_{yb}(y)$ with respect to x and y, respectively, the convective heat transfer coefficient h_k is zero. This results in

$$\phi_{xa}(x) = K\cos(\lambda_{xa}x), \tag{2.22}$$

$$\phi_{yb}(y) = K\cos(\lambda_{yb}y). \tag{2.23}$$

Thus the eigenfunction $\phi_{abc}(x, y, z)$ is given by

$$\phi_{abc}(x, y, z) = K^2\cos(\lambda_{xa}x)\cos(\lambda_{yb}y)\left(K\cos(\lambda_{zc}z) + \frac{h_{\text{down}}}{\lambda_{zc}}\sin(\lambda_{zc}z)\right). \tag{2.24}$$

As derived in [MÖ94, Sec. 3.1] λ_{zc} is a positive scalar which satisfies

$$\frac{K^2\lambda_{zc}^2 - h_{\text{up}}h_{\text{down}}}{K\lambda_{zc}(h_{\text{up}} + h_{\text{down}})} = \cot(\lambda_{zc}L_z), \tag{2.25}$$

The Newton-Raphson method can then be applied to calculate each λ_{zc}, see A.5 for the details. For calculating the eigenvalues λ_{xa} and λ_{yb} the analogous equation of (2.25) is applied. By taking the inverse of (2.25) and taking into account that the convective heat transfer coefficient is zero in x- and y-direction, we have

$$\tan(\lambda_{xa}L_x) = 0 \tag{2.26}$$

$$\tan(\lambda_{yb}L_y) = 0. \tag{2.27}$$

The solution is then given by

$$\lambda_{xa} = \frac{a\pi}{L_x}, \tag{2.28}$$

$$\lambda_{yb} = \frac{b\pi}{L_y}, \tag{2.29}$$

where a, b, c are non-negative integers which are the indices of the eigenvalues and eigenfunctions in x-, y- and z-direction, respectively. λ_{xa}^2, λ_{yb}^2, λ_{zc}^2 are eigenvalues in x-, y- and z-direction, respectively.

In the physical sense $\phi_{abc}(x, y, z)$ represents the abc^{th} eigenmode with respect to the 3D heat transfer system of the die. Its frequencies λ_{xa}, λ_{yb}, λ_{zc} are represented by the eigenvalues in x-, y- and z-direction, respectively. λ_{abc}^2 presents the spectral magnitude of $\phi_{abc}(x, y, z)$.

Remark 2.2. *As the x-, y-, z-directions are orthogonal, the eigenvalues of the system satisfy the equations (2.17) and (2.18). Please refer to A.3 for the proof of orthogonality of the eigenfunctions.*

Since the generated bases $\{\phi_{abc}(x,y,z)\}$ are completely orthogonal in the spatial domain of the die, we introduce the three dimensional finite integral transform with a kernel as $\phi_{abc}(x,y,z)$ [Ölç64], and define the transform $\overline{(\cdot)}_{abc}$ on a function $f(x,y,z,t)$ as

$$\overline{(f)}_{abc}(t) = \iiint_{\Omega} \phi_{abc}(x,y,z) f(x,y,z,t) dxdydz \tag{2.30}$$

where Ω is the whole volume of the die. In the following, the application of the integral transform on a function is indicated by a bar and an index abc. Applying this integral transform (2.30) on $T(x,y,z,t)$ results in

$$\overline{(T)}_{abc}(t) = \iiint_{\Omega} \phi_{abc}(x,y,z) T(x,y,z,t) dxdydz. \tag{2.31}$$

Applying it on $\nabla^2 T(x,y,z,t)$, i.e. $f := \nabla^2 T$, results in

$$\overline{(\nabla^2 T)}_{abc}(t) = \iiint_{\Omega} \phi_{abc}(x,y,z) \nabla^2 T(x,y,z,t) dxdydz. \tag{2.32}$$

According to (2.17), one has

$$\Big(\nabla^2 \phi_{abc}(x,y,z) + \lambda^2 \phi_{abc}(x,y,z)\Big) T(x,y,z,t) = 0 \tag{2.33}$$

Therefore, (2.32) can be rewritten as

$$\begin{aligned}
\overline{(\nabla^2 T)}_{abc}(t) &= \iiint_{\Omega} \Big(\phi_{abc}(x,y,z)\nabla^2 T(x,y,z,t) - (\nabla^2 \phi_{abc}(x,y,z) \\
&\quad + \lambda_{abc}^2 \phi_{abc}(x,y,z)) T(x,y,z,t)\Big) dxdydz \\
&= \iiint_{\Omega} \Big(\phi_{abc}(x,y,z)\nabla^2 T(x,y,z,t) - \nabla^2 \phi_{abc}(x,y,z) T(x,y,z,t)\Big) dxdydz \\
&\quad - \iiint_{\Omega} \lambda_{abc}^2 \phi_{abc}(x,y,z) T(x,y,z,t) dxdydz \\
&= \iiint_{\Omega} \Big(\phi_{abc}(x,y,z)\nabla^2 T(x,y,z,t) - \nabla^2 \phi_{abc}(x,y,z) T(x,y,z,t)\Big) dxdydz \\
&\quad - \lambda_{abc}^2 \overline{(T)}_{abc}(t).
\end{aligned} \tag{2.34}$$

Applying the Gauss Theorem (see Section A.1) in the die region on equation (2.34) (see also [Ölç64]) we get

$$\begin{aligned}
\overline{(\nabla^2 T)}_{abc}(t) = &\sum_{k=1}^{6} \iint_{S_k} \left(\phi_{abc}(x,y,z)\frac{\partial T(x,y,z,t)}{\partial \boldsymbol{n}_k} - T(x,y,z,t)\frac{\partial \phi_{abc}(x,y,z)}{\partial \boldsymbol{n}_k}\right) dS \\
&- \lambda_{abc}^2 \overline{(T)}(\lambda_{abc}, t)
\end{aligned} \tag{2.35}$$

where the point (x, y, z) is on the surface $\mathcal{S}_k$ on the boundary of the die, $\mathcal{S}_k$ is k^{th} surface of the die and $\boldsymbol{n}_k$ is the normal vector on the surface $\mathcal{S}_k$, see also [Ölç64]. Substituting (2.12) and (2.18) into (2.35) results in

$$\overline{(\nabla^2 T)}_{abc}(t) = \sum_{k=1}^{6} \iint_{\mathcal{S}_k} \frac{h_k \phi_{abc}(x,y,z)}{K} T_\infty(x,y,z,t) dS - \lambda_{abc}^2 \overline{(T)}_{abc}(t). \tag{2.36}$$

Applying the integral transform (2.30) on both sides of (2.10) we have

$$\sigma\rho \overline{\left(\frac{\partial T}{\partial t}\right)}_{abc} = K \overline{(\nabla^2 T)}_{abc}(t) + \sum_{i=1}^{N} \overline{(q_i)}_{abc}(t) \tag{2.37}$$

where

$$\overline{(q_i)}_{abc}(t) = \iiint_\Omega \phi_{abc}(x,y,z) q_i(x,y,z,t) dxdydz. \tag{2.38}$$

Substituting (2.36) multiplied by K into (2.37) divided by $\sigma\rho$ results in

$$\frac{d\overline{(T)}_{abc}(t)}{dt} + \lambda_{abc}^2 \frac{K}{\sigma\rho} \overline{(T)}_{abc}(t) = \frac{1}{\sigma\rho} \sum_{i=1}^{N} \overline{(q_i)}_{abc}(t) + \frac{K}{\sigma\rho} \sum_{k=1}^{6} \iint_{\mathcal{S}_k} \frac{h_k \phi_{abc}(x,y,z)}{K} T_\infty(x,y,z,t) dS, \tag{2.39}$$

As the integral transform (2.30) is time independent we have $\overline{\left(\frac{\partial T}{\partial t}\right)}_{abc} = \frac{d\overline{(T)}_{abc}(t)}{dt}$. Therefore, the original system given by (2.10) is transformed to a set of ODEs. The solution of the above equation is

$$\begin{aligned}\overline{(T)}_{abc}(t) =& e^{-\frac{K}{\sigma\rho}\lambda_{abc}^2 t} \overline{(T_0)}_{abc} + \int_0^t e^{\frac{K}{\sigma\rho}\lambda_{abc}^2 (t-\tau)} \left(\frac{1}{\sigma\rho} \sum_{i=1}^{N} \overline{(q_i)}_{abc}(\tau) \right. \\ &\left. + \frac{K}{\sigma\rho} \sum_{k=1}^{6} \iint_{\mathcal{S}_k} \frac{h_k \phi_{abc}(x,y,z)}{K} T_\infty(x,y,z,\tau) dS \right) d\tau,\end{aligned} \tag{2.40}$$

where

$$\overline{(T_0)}_{abc} = \iiint_\Omega \phi_{abc}(x,y,z) T_0(x,y,z) dxdydz, \tag{2.41}$$

and $T_0(x, y, z)$ is the initial temperature of the die at time instant $t = 0$.

By applying the inverse integral transform of (2.30) on (2.40), we have the solution

$$T(x,y,z,t) = \sum_{a=0}^{\infty} \sum_{b=0}^{\infty} \sum_{c=0}^{\infty} G_{abc} \phi_{abc}(x,y,z) \overline{(T)}_{abc}(t), \tag{2.42}$$

where

$$G_{abc} = \frac{1}{\iiint_\Omega \phi_{abc}^2(x,y,z) dxdydz} = \frac{1}{\int_0^{L_x} \phi_{xa}^2(x) dx \int_0^{L_y} \phi_{yb}^2(y) dy \int_0^{L_z} \phi_{zc}^2(z) dz}. \tag{2.43}$$

According to [MÖ94, Sec. 3.1],

$$\int_0^{L_z} \phi_{zc}^2(z)dz = \frac{1}{2}\frac{(h_{\text{down}}^2 + K^2\lambda_{zc}^2)(\frac{Kh_{\text{up}}}{h_{\text{up}}+K^2\lambda_{zc}^2} + L_z) + Kh_{\text{down}}}{\lambda_{zc}^2}. \tag{2.44}$$

For the value of $\int_0^{L_x} \phi_{xa}^2(x)dx$ and $\int_0^{L_y} \phi_{yb}^2(y)dy$ with respect to x and y respectively the convective heat transfer coefficient h_k is zero. This result in

$$\int_0^{L_x} \phi_{xa}^2(x)dx = \frac{1}{2}K^2 L_x \tag{2.45}$$

and

$$\int_0^{L_y} \phi_{yb}^2(y)dy = \frac{1}{2}K^2 L_y \tag{2.46}$$

However, when $a = 0$, $\phi_{xa} = K\cos(\frac{a\pi}{L_x}x) = K$, which results in

$$\int_0^{L_x} \phi_{xa}^2(x)dx = K^2 L_x. \tag{2.47}$$

When $b = 0$, $\phi_{yb} = K\cos(\frac{b\pi}{L_y}y) = K$, which results in

$$\int_0^{L_y} \phi_{yb}^2(y)dy = K^2 L_y \tag{2.48}$$

Remark 2.3. *The solution given in (2.42) requires an infinite summation. However, in practical application we only need to sum up the terms with dominating eigenvalues, see for example [HL09] and the references therein.*

Remark 2.4. *In equations* (2.21) *and* (2.44), h_{up} *and* h_{down} *appear asymmetrical. This is caused by the way of solving the eigenvalue system [MÖ94]. In the given derivation the eigenfunction is* (2.21), *and the normalized eigenfunction is*

$$\frac{1}{Z_c}\big(K\cos(\lambda_{zc}z) + \frac{h_{\text{down}}}{\lambda_{zc}}\sin(\lambda_{zc}z)\big)$$

with the norm $Z_c = \sqrt{\int_0^{L_z} \phi_{zc}^2(z)dz}$. *If the coordinated system is defined with an opposed z-direction, see Figure 2.5, the* h_{up} *and* h_{down} *will be exchanged in the derivations. However, the solution of the eigenvalues, the normalized eigenfunctions, and the temperature distribution is equivalent.*

2.4 Example for the 3D PDE heat transfer model

To verify the proposed 3D thermal behavior model, we consider the 8-core IBM Cell Processor as an example. The floorplan of the architecture is shown in Figure 2.6. It

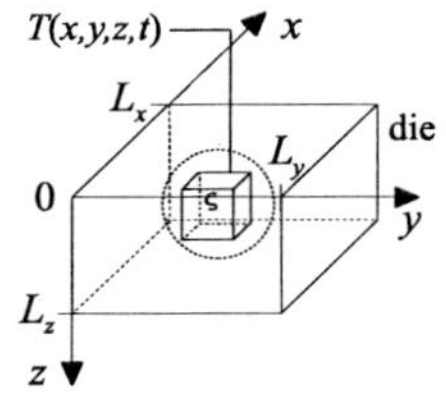

Figure 2.5: Coordiate system with opposed z-direction

consists of the Power Processor Element (PPE) and eight Synergistic Processor Elements (SPEs) each with its own local memory (LS). The processor has a 500M L2 cache, a high bandwidth internal Element Interconnect Bus (EIB), two configurable non-coherent I/O interfaces and a Memory Interface Controller (MIC). It is produced with a 90 nm technology node.

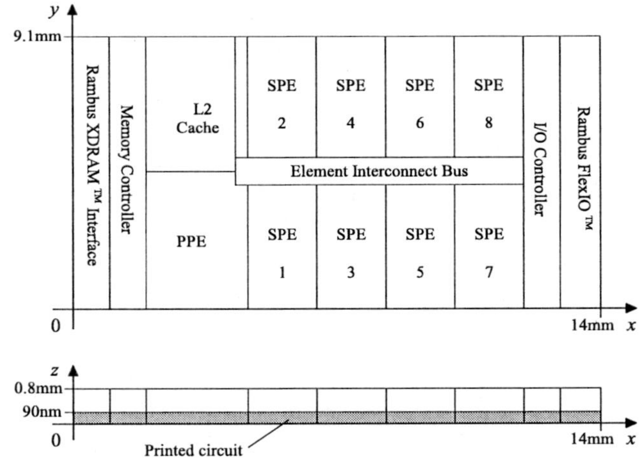

Figure 2.6: The floorplan and side view of the IBM Cell

Remark 2.5. ***Technology node*** *is defined by the smallest printed feature, which is the measurement of the technological advancement of the microelectronic chips. For micro processing units, this value refers to the half-pitch of the polysilicon lines, or to the printed gate length [DW04].*

In this simulation, it is assumed that the voltage can be adjusted continuously. The parameters are chosen as $\sigma = 707\,\mathrm{J/(kg \cdot K)}$, $\rho = 2330\,\mathrm{kg/m^3}$, $K = 100\,\mathrm{W/(m \cdot K)}$, $h_{\text{up}} = 1000\,\mathrm{W/(m^2 \cdot K)}$ and $h_{\text{down}} = 100\,\mathrm{W/(m^2 \cdot K)}$ [Geb71, Hol86, BILD90, MSS+05]. Assume that the environment temperature is 298 K and the chip initial temperature is $T_0(x, y, z) = 298\,\mathrm{K}$. Suppose that the heat source has a uniform distribution in the printed circuit volume, i.e.

$$q_i(x, y, z, t) = \frac{P_i(t)}{V_{\mathrm{P},i}}, \tag{2.49}$$

where $V_{\mathrm{P},i}$ is the printed circuit volume of the core i, (x, y, z) is a point in the area of the printed circuit in core i and $P_i(t)$ is the consumed power of the core i.

A constant steplike power $P_i(t)$ is applied to each core and the PPE. In this simulation $q_i(x, y, z, t) = 1.55 \cdot 10^{12}\,\mathrm{W/m^3}$ for all points (x, y, z) in the printed circuit volume of the cores and the PPE. As explained in Remark 2.3 we only sum up the terms with eigenvalues to obtain the temperature distribution by applying equation (2.42). The stationary value of the step response on a fixed z-plane at $z = 0.5\,\mathrm{mm}$ for a given number of eigenvalues are shown in Figure 2.7 - 2.9. The Figures 2.7 - 2.9 show the temperature distribution with different calculation accuracy. The distribution in Figure 2.7 with 50 eigenvalues differs from Figure 2.8 and 2.9. Figure 2.8 and 2.9 show a highly similar distribution. The number of eigenvalues to solve the PDE equation needs to be chosen large enough to achieve an accurate solution, see Remark 2.6 for finding a suitable number eigenvalues. In this example, a highly precise solution can be achieved with 800 eigenvalues. Besides, two points in core 2 and core 3 are selected under the condition that all the cores work equally. Figure 2.10 shows the dynamical step response of the two selected points of the die.

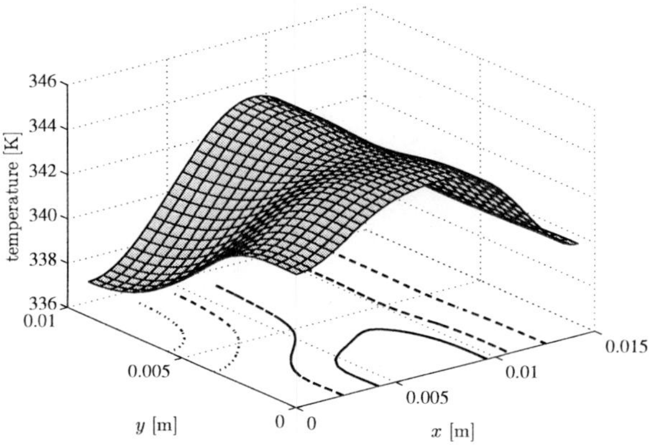

Figure 2.7: Temperature distribution with 50 eigenvalues

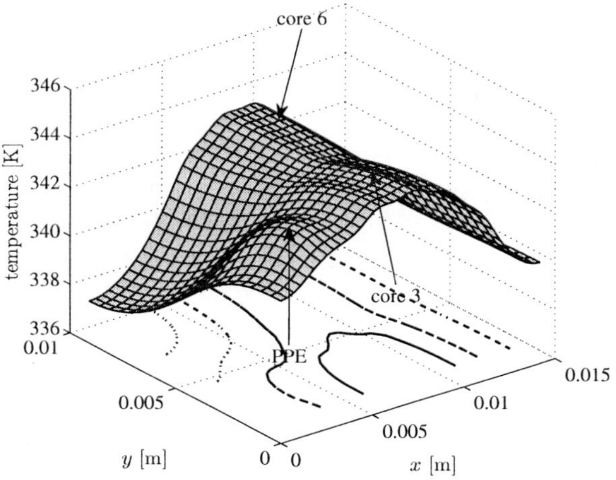

Figure 2.8: Temperature distribution with 800 eigenvalues

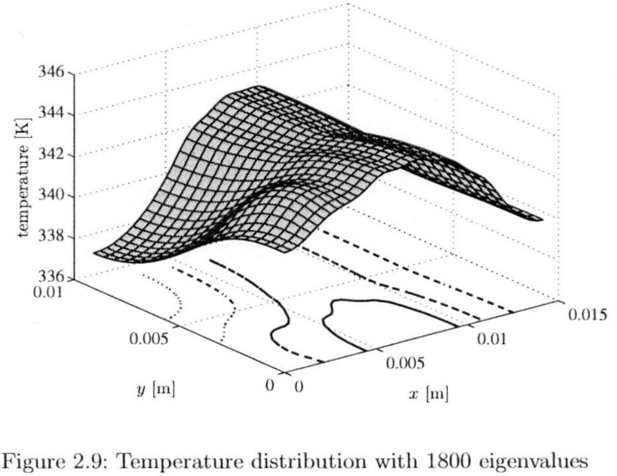

Figure 2.9: Temperature distribution with 1800 eigenvalues

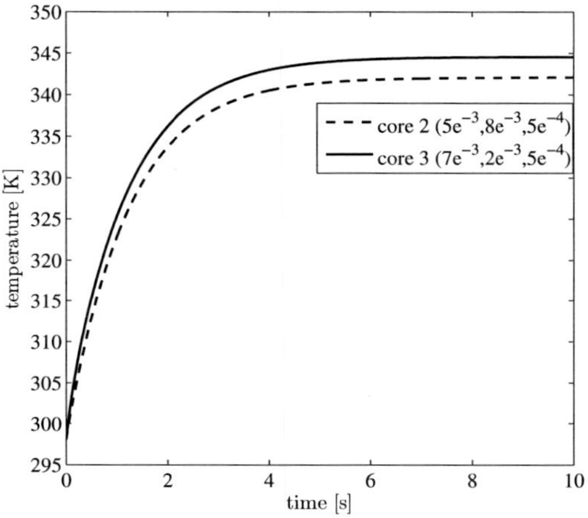

Figure 2.10: The step dynamical response of two points

Remark 2.6. *As discussed above, the computational accuracy of the temperature distribution can be controlled by choosing the number of the eigenvalues. An error function can be chosen to measure the computational accuracy. A possible form of the error function is the 2-Norm of the distribution difference between the eigenvalue number n and $n+1$. A threshold can be set, such that when the error is smaller than the threshold, the computation can be stopped.*

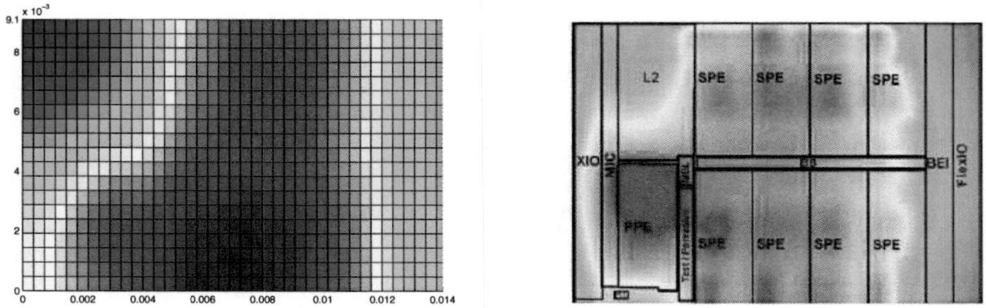

Figure 2.11: Comparison of the simulation result from Figure 2.8 (left) and the one from [PBB+05] (right)

Figure 2.11 shows the presented simulation result under 800 eigenvalues in comparison with the simulation results presented in [PBB+05] both for an 8-core IBM CELL processor. In principle the heat distribution is in both cases similar. However, unfortunately [PBB+05] does not present any information on how the simulation is conducted and under which operation conditions. Therefore, some differences between the two simulation results appear.

In order to show the cross thermal effect among the cores, the same step input is given to core 2, core 5 and the PPE. The other cores are set as idle, i.e. $q_i(x, y, z, t) = 0$. The simulation result given in Figure 2.12 shows that core 2, core 5 and the PPE have a higher temperature than the other parts. Meanwhile, the whole die is heated because of the heat conduction in the die.

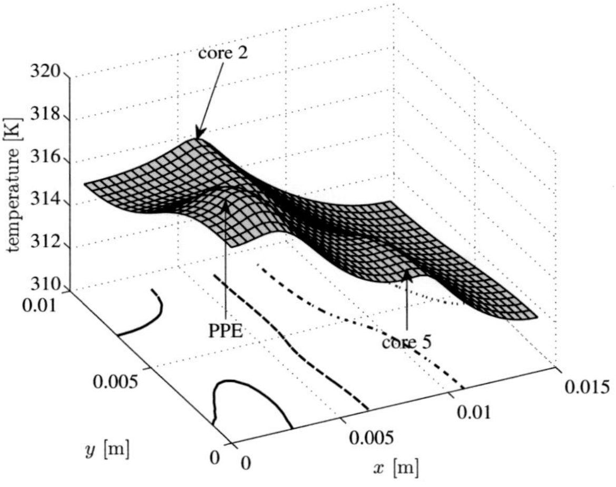

Figure 2.12: Temperature distribution with partial load

2.5 Summary

In this chapter, the multi-core processor is introduced and the power consumption is analyzed. The general 3D heat conduction PDE model is given. The eigenfunction and eigenvalue system based PDE solution algorithm is employed to solve the proposed PDE. An IBM CELL based example is presented. In the following chapter, the thermal and power balancing/management policy is developed based on the derived mode.

3 Thermal and power balancing/management policy design for 2D MCP

As mentioned in the first chapter, the temperature of the die has significant influence on the chip's operation reliability and lifespan. According to [VWWL00], an increase of the average operating temperature of 10-15 K can cause a two times reduction of the chip lifespan. In the model of the chip thermal behavior presented in Chapter 2 the die is considered as an integral system, i.e. the dynamics of the cores are not modeled individually. However, the objective is to balance the temperature and the power dissipation of each core. As each core performs an individual temperature distribution, we need to define a criterion to measure the temperature variation among the cores. Hence, we consider each core as an individual system and investigate the heat exchange among the cores. Based on this, a 1D PDE can be introduced for each cores and the average temperature of each core on a fixed plane is defined as the new state. Within this model, the thermal cross influence among the cores can be described clearly.

The following assumptions are introduced for this chapter. First, we assume that at least one digital thermal sensor is placed at each core and there is a thermal management center to monitor the temperature of each core. Furthermore, the initial temperature of the whole die satisfies $T_0(x, y, z, 0) \equiv T_\infty$.

In this Chapter, based on the energy conservation law, the 3D model is transformed into a model described by a group of 1D PDEs. A quadratic cost function which contains the control input and the temperature difference among the cores is introduced. The Riccati equation approach is employed to obtain the controller.

3.1 Transformation from the 3D model to the 1D model

Definition 3.1. *Define* $T_{\mathrm{a},i}(z,t)$ [K] *as the average over-temperature on a fixed* z *plane*

$$T_{\mathrm{a},i}(z,t) = \frac{\int_{x_{i1}}^{x_{i2}} \int_{y_{i1}}^{y_{i2}} (T(x,y,z,t) - T_\infty) dx dy}{S_i} \quad \forall i \in \{1, 2, ..., N\}, \tag{3.1}$$

where $S_i = (x_{i2} - x_{i1})(y_{i2} - y_{i1})$ *and* x_{i1}, x_{i2}, y_{i1} *and* y_{i2} *are the boundary coordinate values in* x*- and* y*-direction of core* i*.*

Suppose the thermal sensors are located on top of the die and sample the average temperature of each core. Thus, the measured output of core i is

$$\xi_{1,i}(t) = T_{a,i}(L_z, t), \tag{3.2}$$

where L_z is the size of the die in z-direction as shown in Figure 3.1.

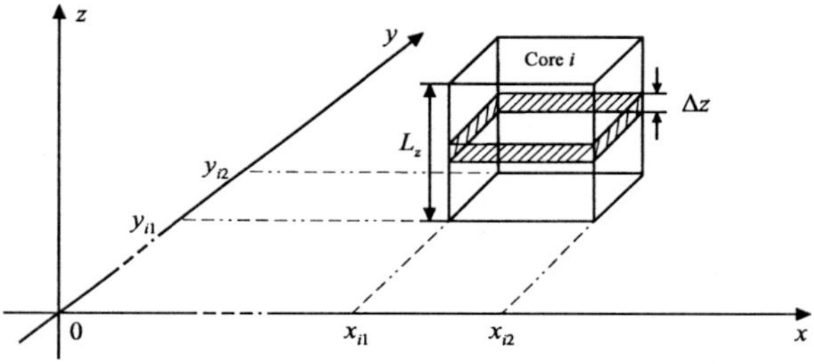

Figure 3.1: Transformation from 3D PDE to 1D PDEs

In order to transform the 3D PDE to a group of 1D PDEs, we consider the heat conduction schematic diagram shown in Figure 3.2.

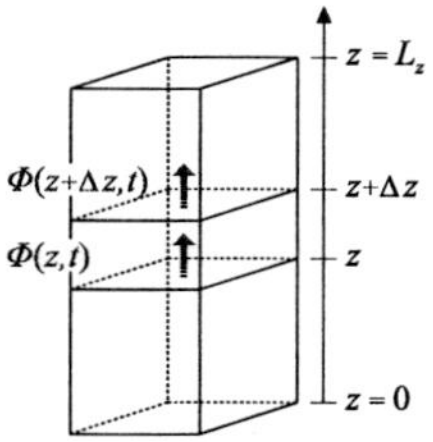

Figure 3.2: 1D heat conduction schematic diagram

Considering an infinitesimal volume between z and $z + \Delta z$, see Figure 3.2, the time rate of the heat transfer in this volume is

$$\frac{dQ_i(z,t)}{dt} = \sigma \rho S_i \Delta z \frac{\partial T_{a,i}(z,t)}{\partial t}. \tag{3.3}$$

where σ [J/(kg · K)] represents the specific heat capacity of the die and ρ [kg/m^3] represents the density of the die as explained in Chapter 2.

According to the energy conservation law, the heat variation in the volume $S_i \Delta z$ consists of three parts, the heat conduction inside the core (this represents the heat transfer in z-direction), the heat exchanged with the outside (i.e. in x- and y-direction), and the internal heat source

$$\frac{dQ_i(z,t)}{dt} = \frac{dQ_{\text{con},i}(z,t)}{dt} + \frac{dQ_{\text{ex},i}(z,t)}{dt} + \frac{dQ_{\text{in},i}(z,t)}{dt}. \tag{3.4}$$

In the right hand side of equation (3.4), the first term is the time rate of the heat conduction in this core inside, the second term is the heat exchange rate between this core and its outside, while the third term is the heat variation due to the internal heat source. The first term can be calculated by

$$\frac{dQ_{\text{con},i}(z,t)}{dt} = \Phi_i(z,t)S_i - \Phi_i(z+\Delta z,t)S_i, \tag{3.5}$$

where $\Phi_i(z,t)$ [W/m^2] is the heat flux in z-direction of core i with the same positive direction as the z-axis. The heat flux represents the heat energy transfer rate through a surface. According to the Fourier heat conduction law [Fou09], one has

$$\Phi_i(z,t) = -K\frac{\partial T_{\text{a},i}(z,t)}{\partial z}. \tag{3.6}$$

with the thermal conductivity K [W/(m · K)] of the core. Equation (3.6) follows from (2.6) by setting $\boldsymbol{n} = (0\ 0\ 1)^T$ with respect to coordinate system defined in Figure 3.1. As Δz is infinitesimal we have

$$\begin{aligned}\Phi_i(z,t) - \Phi_i(z+\Delta z,t) =& K\frac{\partial T_{\text{a},i}(z+\Delta z,t)}{\partial z} - K\frac{\partial T_{\text{a},i}(z,t)}{\partial z}\\ =& K\frac{\partial^2 T_{\text{a},i}(z,t)}{\partial z^2}\Delta z.\end{aligned} \tag{3.7}$$

Substituting (3.7) in (3.5) yields

$$\frac{dQ_{\text{con},i}(z,t)}{dt} = S_i \Delta z K\frac{\partial^2 T_{\text{a},i}(z,t)}{\partial z^2}. \tag{3.8}$$

The heat exchange rate of (3.4) is

$$\begin{aligned}\frac{dQ_{\text{ex},i}(z,t)}{dt} =& \Bigg(\int_{x_{i1}}^{x_{i2}} \Phi_{\text{ex}}(x,y,z,t)dx\bigg|_{y=y_{i1}} + \int_{x_{i1}}^{x_{i2}} \Phi_{\text{ex}}(x,y,z,t)dx\bigg|_{y=y_{i2}}\\ &+ \int_{y_{i1}}^{y_{i2}} \Phi_{\text{ex}}(x,y,z,t)dy\bigg|_{x=x_{i1}} + \int_{y_{i1}}^{y_{i2}} \Phi_{\text{ex}}(x,y,z,t)dy\bigg|_{x=x_{i2}}\Bigg)\Delta z,\end{aligned} \tag{3.9}$$

where $\Phi_{\mathrm{ex}}(x,y,z,t)$ [W/m^2] is the heat flux through the boundaries of the core in x- and y-direction. According to Fourier's heat conduction law [Fou09]

$$\Phi_{\mathrm{ex}}(x,y,z,t) = -K\frac{\partial T(x,y,z,t)}{\partial \boldsymbol{n}} \tag{3.10}$$

where $\boldsymbol{n}$ is normal vector on the boundary areas of the core, i.e. in the positive and negative x- and y-direction respectively, see also Figure 3.1.

In the following, the exchanged heat through the boundaries which is caused by the power consumption of the cores will be interpreted as an energy generation function per time and volume unit resulting in

$$\frac{dQ_{\mathrm{ex},i}(z,t)}{dt} = S_i\Delta z\sum_{j=1}^{N} q_{\mathrm{ex},ij}(z,P_j,t). \tag{3.11}$$

where P_j [W] is the power consumption of core j and $q_{\mathrm{ex},ij}(z,P_j,t)$ [W/m^3] is the energy generation function due to the heat exchange of core i with the outside caused by the power consumption of core j. Setting (3.9) and (3.11) equal and substituting (3.10) into it yields

$$\begin{aligned}\sum_{j=1}^{N} q_{\mathrm{ex},ij}(z,P_j,t) =&\frac{K}{S_i}\Bigg(-\int_{x_{i1}}^{x_{i2}}\frac{\partial T(x,y,z,t)}{\partial y}dx\Bigg|_{y=y_{i1}} + \int_{x_{i1}}^{x_{i2}}\frac{\partial T(x,y,z,t)}{\partial y}dx\Bigg|_{y=y_{i2}}\\ &-\int_{y_{i1}}^{y_{i2}}\frac{\partial T(x,y,z,t)}{\partial x}dy\Bigg|_{x=x_{i1}} + \int_{y_{i1}}^{y_{i2}}\frac{\partial T(x,y,z,t)}{\partial x}dy\Bigg|_{x=x_{i2}}\Bigg).\end{aligned} \tag{3.12}$$

Substituting (2.42), which is the solution of the PDE introduced in Section 2.2 (i.e. a temperature distribution [K]) into (3.12), results in

$$\begin{aligned}q_{\mathrm{ex},ij}(z,P_j,t) =&\frac{K}{S_i}\Bigg(-\int_{x_{i1}}^{x_{i2}}\sum_{a=0}^{\infty}\sum_{b=0}^{\infty}\sum_{c=0}^{\infty}G_{abc}\frac{\partial\phi_{abc}}{\partial y}\overline{(T)}_{abc,j}(t)dx\Bigg|_{y=y_{i1}}\\ &+\int_{x_{i1}}^{x_{i2}}\sum_{a=0}^{\infty}\sum_{b=0}^{\infty}\sum_{c=0}^{\infty}G_{abc}\frac{\partial\phi_{abc}}{\partial y}\overline{(T)}_{abc,j}(t)dx\Bigg|_{y=y_{i2}}\\ &-\int_{y_{i1}}^{y_{i2}}\sum_{a=0}^{\infty}\sum_{b=0}^{\infty}\sum_{c=0}^{\infty}G_{abc}\frac{\partial\phi_{abc}}{\partial x}\overline{(T)}_{abc,j}(t)dy\Bigg|_{x=x_{i1}}\\ &+\int_{y_{i1}}^{y_{i2}}\sum_{a=0}^{\infty}\sum_{b=0}^{\infty}\sum_{c=0}^{\infty}G_{abc}\frac{\partial\phi_{abc}}{\partial x}\overline{(T)}_{abc,j}(t)dy\Bigg|_{x=x_{i2}}\Bigg),\end{aligned} \tag{3.13}$$

with

$$\overline{(T)}_{abc,j}(t) = \int_0^t e^{\frac{K}{\sigma\rho}\lambda_{abc}^2(t-\tau)}\frac{1}{\sigma\rho}\overline{q}_{abc,j}(\tau)d\tau. \tag{3.14}$$

The heat variation due to the internal heat source is

$$\frac{dQ_{\text{in},i}(z,t)}{dt} = S_i \Delta z q_i(z,t) \tag{3.15}$$

with

$$q_i(z,t) = \frac{\int_{x_{i1}}^{x_{i2}} \int_{y_{i1}}^{y_{i2}} q_i(x,y,z,t)dxdy}{S_i} \tag{3.16}$$

where $q_i(x,y,z,t)$ [W/m^3] is the internal heat generation function per time unit and per volume unit of core i, see also Section 2.2.

Substituting (3.3), (3.8), (3.11) and (3.15) into (3.4) and dividing it by $S_i\Delta z$ results in

$$\sigma\rho\frac{\partial T_{\text{a},i}(z,t)}{\partial t} = K\frac{\partial^2 T_{\text{a},i}(z,t)}{\partial z^2} + \sum_{j=1}^{N} q_{\text{ex},ij}(z,P_j,t) + q_i(z,t), \tag{3.17}$$

with the boundary conditions

$$K\frac{\partial T_{\text{a},i}(L_z,t)}{\partial z} = -\, h_{\text{up}} T_{\text{a},i}(L_z,t), \tag{3.18a}$$

$$K\frac{\partial T_{\text{a},i}(0,t)}{\partial z} = h_{\text{down}} T_{\text{a},i}(0,t). \tag{3.18b}$$

The equations (3.17)-(3.18) represent the 1D PDE describing the average over-temperature dynamics.

For controlling the temperature balance the separation of variables is applied on the heat generation function $q_{\text{ex},ij}(z,P_j,t)$. It can be approximated by considering its influence within the range of the time constant t_{c}

$$q_{\text{ex},ij}(z,P_j,t) \approx \sum_{l=1}^{L_{\text{d}}} \kappa_{ilj}(z) P_j(t - l\cdot t_{\text{d}}) \tag{3.19}$$

$$t_{\text{d}} = t_{\text{c}}/L_{\text{d}} \tag{3.20}$$

where $L_{\text{d}} \in \mathbb{N}$ is a constant relative to the calculation accuracy. The variable κ_l can be obtained from (3.13) by replacing $\overline{(T)}_{abc,j}(t)$ by

$$\overline{(T)}_{abc,j,l}(t) = \int_{t-l\cdot t_{\text{d}}}^{t} e^{\frac{K}{\sigma\rho}\lambda^2_{abc}(t-\tau)} \frac{1}{\sigma\rho} d\tau \overline{(q)}_{abc,i}(t - l\cdot t_{\text{d}}). \tag{3.21}$$

Remark 3.1. *The time constant t_{c} is derived from the system* (2.10) *with the boundary conditions* (2.13)-(2.16) *of the 1D model proposed in this chapter. The constant t_{c} can be gotten from the first eigenvalue, which is λ_{000}, as this is the eigenvalue relative to the slowest dynamical response.*

Based on (3.19) we have

$$\sum_{j=1}^{N} q_{\mathrm{ex},ij}(z,P_j,t) \approx \sum_{l=1}^{L_\mathrm{d}} \boldsymbol{\kappa}_{il}(z)\boldsymbol{P}(t-l\cdot t_\mathrm{d}) \tag{3.22}$$

with

$$\boldsymbol{\kappa}_{il}(z) = \Big[\kappa_{il1},\, \dots,\, \kappa_{ilN}\Big] \tag{3.23}$$

$$\boldsymbol{P}(t-l\cdot t_\mathrm{d}) = \Big[P_1(t-l\cdot t_\mathrm{d}),\, \dots,\, P_N(t-l\cdot t_\mathrm{d})\Big]^T \tag{3.24}$$

Further, the internal heat generation function is given by

$$q_i(z,t) = \kappa_{i0i}P_i(t) \tag{3.25}$$

with κ_{i0i} according to (2.49). Setting $\kappa_{i0j} = 0$ for all $j \neq i$ the system (3.17)-(3.18) can be written in a symbolic way in the state space. The state is the trajectory segment $T_{\mathrm{a},i}(\cdot,t) = \{T_{\mathrm{a},i}(z,t), 0 \le z \le L_z\}$ [CZ95, Sec. 2.1]. Define the operators $\boldsymbol{A}_i$ and $\boldsymbol{B}_{i,l}$ as the linear continuous mappings

$$\underbrace{\begin{bmatrix} \dot{T}_{\mathrm{a},i}(0,t) \\ \dot{T}_{\mathrm{a},i}(z,t) \\ \dot{T}_{\mathrm{a},i}(L_z,t) \end{bmatrix}}_{\dot{\boldsymbol{T}}_{\mathrm{a},i}(t)} = \underbrace{\begin{bmatrix} \frac{h_2}{K}\frac{\partial}{\partial z} & 0 & 0 \\ 0 & \frac{K}{\sigma\rho}\frac{\partial^2}{\partial z^2} & 0 \\ 0 & 0 & -\frac{h_1}{K}\frac{\partial}{\partial z} \end{bmatrix}}_{\boldsymbol{A}_i} \underbrace{\begin{bmatrix} T_{\mathrm{a},i}(0,t) \\ T_{\mathrm{a},i}(z,t) \\ T_{\mathrm{a},i}(L_z,t) \end{bmatrix}}_{\boldsymbol{T}_{\mathrm{a},i}(t)} + \sum_{l=0}^{L_\mathrm{d}} \underbrace{\begin{bmatrix} 0 \\ \frac{\boldsymbol{\kappa}_{il}(z)}{\sigma\rho} \\ 0 \end{bmatrix}}_{\boldsymbol{B}_{i,l}} \boldsymbol{P}(t-l\cdot t_\mathrm{d}) \tag{3.26}$$

Therefore, the system can be described as

$$\dot{\boldsymbol{T}}_{\mathrm{a},i}(t) = \boldsymbol{A}_i\boldsymbol{T}_{\mathrm{a},i}(t) + \sum_{l=0}^{L_\mathrm{d}} \boldsymbol{B}_{i,l}\boldsymbol{P}(t-l\cdot t_\mathrm{d}) \tag{3.27}$$

Here $\boldsymbol{A}_i$ is an infinitesimal operator (as defined in [CZ95, Sec. 2.1], for details see A.6), which describes the state operator of the core i, and $\boldsymbol{A}_i$ describes the temperature variety according to internal heat conduction and heat convection with the environment. $\boldsymbol{B}_{i,l} \in \mathscr{L}(\boldsymbol{U}, \mathscr{T}_\mathrm{a})$ is the input operator, which represents the temperature influence by the power consumption of the cores, where $\boldsymbol{U}, \mathscr{T}_\mathrm{a}$ are Hilbert spaces which indicate the input and state space and $\mathscr{L}(\boldsymbol{U}, \mathscr{T}_\mathrm{a})$ is the bounded linear operator from Hilbert space $\boldsymbol{U}$ to the Hilbert space $\mathscr{T}_\mathrm{a}$.

3.2 Optimal thermal balance policy

To balance the temperature difference among the cores, we define the regulated output as

$$\boldsymbol{\xi}_2(z,t) = \Big[\xi_{2,1}(z,t), \xi_{2,2}(z,t),\, \dots,\, \xi_{2,N}(z,t)\Big]^T, \tag{3.28}$$

with

$$\xi_{2,i}(z,t) = T_{\mathrm{a},i}(z,t) - \frac{1}{N}\sum_{j=1}^{N} T_{\mathrm{a},j}(z,t), \tag{3.29}$$

which is the temperature difference between the average temperature of all cores and the temperature of core i on a fixed z-plane.

Defining

$$\boldsymbol{T}_{\mathrm{a}}(t) = \begin{bmatrix} \boldsymbol{T}_{\mathrm{a},1}(t) \\ \vdots \\ \boldsymbol{T}_{\mathrm{a},N}(t) \end{bmatrix}, \quad \boldsymbol{A} = \begin{bmatrix} \boldsymbol{A}_1 & & \\ & \ddots & \\ & & \boldsymbol{A}_N \end{bmatrix}, \quad \boldsymbol{B}_l = \begin{bmatrix} \boldsymbol{B}_{l,1} \\ \vdots \\ \boldsymbol{B}_{l,N} \end{bmatrix}, \tag{3.30}$$

the infinite-dimensional system is given by

$$\dot{\boldsymbol{T}}_{\mathrm{a}}(t) = \boldsymbol{A}\boldsymbol{T}_{\mathrm{a}} + \sum_{l=0}^{L_{\mathrm{d}}} \boldsymbol{B}_l \boldsymbol{P}(t - l \cdot t_{\mathrm{d}}) \tag{3.31a}$$

$$\boldsymbol{\xi}_1(t) = \mathscr{C}_1 \boldsymbol{T}_{\mathrm{a}}(t) \tag{3.31b}$$

$$\boldsymbol{\xi}_2(t) = \mathscr{C}_2 \boldsymbol{T}_{\mathrm{a}}(t) \tag{3.31c}$$

with $\mathscr{C}_1 \in \mathscr{L}(\mathscr{T}_{\mathrm{a}}, \Xi_1)$ and $\mathscr{C}_2 \in \mathscr{L}(\mathscr{T}_{\mathrm{a}}, \Xi_2)$ where Ξ_1 is the measurement output space and Ξ_2 is regulated output space.

As shown in (3.22), we employ the numeral approximation to calculate the temperature distribution. To get higher calculation accuracy, larger L_{d} should be adopted. For the control design the system (3.31a) is transformed into a system without input delay, see [ZTH09]. The solution of (3.31a) is given by

$$\boldsymbol{T}_{\mathrm{a}}(t) = \boldsymbol{T}_{\mathrm{a}}(t_0) e^{\boldsymbol{A}(t-t_0)} + \sum_{l=0}^{L_{\mathrm{d}}} \int_{t_0}^{t} e^{\boldsymbol{A}(t-\tau)} \boldsymbol{B}_l \boldsymbol{P}(\tau - l \cdot t_{\mathrm{d}}) d\tau. \tag{3.32}$$

Substituting $\tau' = \tau - l \cdot t_{\mathrm{d}}$ we have

$$\boldsymbol{T}_{\mathrm{a}}(t) = \boldsymbol{T}_{\mathrm{a}}(t_0) e^{\boldsymbol{A}(t-t_0)} + \sum_{l=0}^{L_{\mathrm{d}}} \int_{t_0 - l\cdot t_{\mathrm{d}}}^{t - l\cdot t_{\mathrm{d}}} e^{\boldsymbol{A}(t-\tau'-l\cdot t_{\mathrm{d}})} \boldsymbol{B}_l \boldsymbol{P}(\tau') d\tau' \tag{3.33}$$

which can equivalently be written as

$$\begin{aligned} \boldsymbol{T}_{\mathrm{a}}(t) = \boldsymbol{T}_{\mathrm{a}}(t_0) e^{\boldsymbol{A}(t-t_0)} &+ \sum_{l=0}^{L_{\mathrm{d}}} \int_{t_0 - l\cdot t_{\mathrm{d}}}^{t_0} e^{\boldsymbol{A}(t-\tau'-l\cdot t_{\mathrm{d}})} \boldsymbol{B}_l \boldsymbol{P}(\tau') d\tau' \\ &+ \sum_{l=0}^{L_{\mathrm{d}}} \int_{t_0}^{t - l\cdot t_{\mathrm{d}}} e^{\boldsymbol{A}(t-\tau'-l\cdot t_{\mathrm{d}})} \boldsymbol{B}_l \boldsymbol{P}(\tau') d\tau' \end{aligned} \tag{3.34}$$

where the first integral on the right hand side represents the influence of the past input before the initial time t_0. Equation (3.34) can also be written as

$$\begin{aligned}\boldsymbol{T}_{\mathrm{a}}(t) = &\left(\boldsymbol{T}_{\mathrm{a}}(t_0) + \sum_{l=0}^{L_{\mathrm{d}}} \int_{t_0 - l\cdot t_{\mathrm{d}}}^{t_0} e^{\boldsymbol{A}(t_0-\tau'-l\cdot t_{\mathrm{d}})} \boldsymbol{B}_l \boldsymbol{P}(\tau')d\tau'\right) e^{\boldsymbol{A}(t-t_0)} \\ &+ \sum_{l=0}^{L_{\mathrm{d}}} \int_{t_0}^{t} e^{\boldsymbol{A}(t-\tau')} e^{-\boldsymbol{A}(l\cdot t_{\mathrm{d}})} \boldsymbol{B}_l \boldsymbol{P}(\tau')d\tau' - \sum_{l=0}^{L_{\mathrm{d}}} \int_{t-l\cdot t_{\mathrm{d}}}^{t} e^{\boldsymbol{A}(t-\tau')} e^{-\boldsymbol{A}(l\cdot t_{\mathrm{d}})} \boldsymbol{B}_l \boldsymbol{P}(\tau')d\tau'\end{aligned} \tag{3.35}$$

Defining

$$\widehat{\boldsymbol{T}}_{\mathrm{a}}(t) = \boldsymbol{T}_{\mathrm{a}}(t) + \sum_{l=0}^{L_{\mathrm{d}}} \int_{t-l\cdot t_{\mathrm{d}}}^{t} e^{\boldsymbol{A}(t-\tau)} e^{-\boldsymbol{A}(l\cdot t_{\mathrm{d}})} \boldsymbol{B}_l \boldsymbol{P}(\tau)d\tau. \tag{3.36}$$

with

$$\widehat{\boldsymbol{T}}_{\mathrm{a}}(t_0) = \boldsymbol{T}_{\mathrm{a}}(t_0) + \sum_{l=0}^{L_{\mathrm{d}}} \int_{t_0-l\cdot t_{\mathrm{d}}}^{t_0} e^{\boldsymbol{A}(t_0-\tau-l\cdot t_{\mathrm{d}})} \boldsymbol{B}_l \boldsymbol{P}(\tau)d\tau \tag{3.37}$$

the dynamics of $\widehat{\boldsymbol{T}}_{\mathrm{a}}(t)$ are given by

$$\dot{\widehat{\boldsymbol{T}}}_{\mathrm{a}}(t) = \boldsymbol{A}\widehat{\boldsymbol{T}}_{\mathrm{a}}(t) + \widehat{\boldsymbol{B}}\boldsymbol{P}(t), \tag{3.38}$$

where $\widehat{\boldsymbol{B}} = \sum_{l=0}^{L_{\mathrm{d}}} e^{-\boldsymbol{A}(l\cdot t_{\mathrm{d}})}\boldsymbol{B}_l$. Thus the original system (3.31a) is transformed into a non-delayed 1D PDE system. Systems (3.27) and (3.38) have the same eigenvalues and eigenfunctions. To meet the intended workload target of the processor, the target average frequency is set as $f_{\mathrm{t}}(t)$ [ZAD09], the corresponding supply voltage is $V_{\mathrm{ddt}}(t)$ for each core, and suppose $\boldsymbol{P}_{\mathrm{t}}(t) = \boldsymbol{P}(f_{\mathrm{t}}(t), V_{\mathrm{ddt}}(t))$, which has the form

$$\boldsymbol{P}_{\mathrm{t}}(t) = \left[P_{\mathrm{t},1}(t), P_{\mathrm{t},2}(t), ..., P_{\mathrm{t},N}\right]^T. \tag{3.39}$$

The power input $\boldsymbol{P}(t)$ consists of the target power $\boldsymbol{P}_{\mathrm{t}}(t)$ and the control input $\boldsymbol{P}_{\mathrm{c}}(t)$, i.e. $\boldsymbol{P}(t) = \boldsymbol{P}_{\mathrm{t}}(t) + \boldsymbol{P}_{\mathrm{c}}(t)$. Therefore we have an affine system

$$\dot{\widehat{\boldsymbol{T}}}_{\mathrm{a}}(t) = \boldsymbol{A}\widehat{\boldsymbol{T}}_{\mathrm{a}}(t) + \widehat{\boldsymbol{B}}\boldsymbol{P}_{\mathrm{c}}(t) + \widehat{\boldsymbol{B}}\boldsymbol{P}_{\mathrm{t}}(t). \tag{3.40}$$

In order to reach the task target, the target input $\boldsymbol{P}_{\mathrm{t}}(t)$ is not considered in the control design, i.e. we consider the system

$$\dot{\widehat{\boldsymbol{T}}}_{\mathrm{a}}(t) = \boldsymbol{A}\widehat{\boldsymbol{T}}_{\mathrm{a}}(t) + \widehat{\boldsymbol{B}}\boldsymbol{P}_{\mathrm{c}}(t). \tag{3.41}$$

Further, the regulated output is substituted according to

$$\widehat{\boldsymbol{\xi}}_2(t) = \mathscr{C}_2\widehat{\boldsymbol{T}}_{\mathrm{a}}(t). \tag{3.42}$$

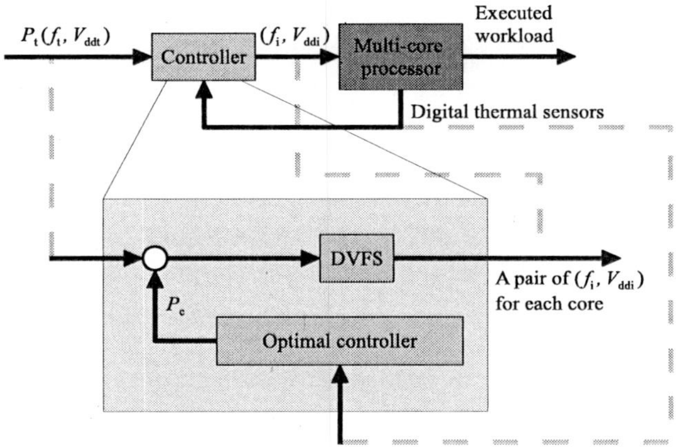

Figure 3.3: Thermal balance controller structure

The thermal balance controller structure is shown in Figure 3.3. Define the cost function

$$J = \int_0^{\infty} \Big[< \boldsymbol{P}_c(t), \mathscr{R}\boldsymbol{P}_c(t) > + < \widehat{\boldsymbol{\xi}}_2(t), \mathscr{Q}\widehat{\boldsymbol{\xi}}_2(t) > \Big] dt, \tag{3.43}$$

where $\mathscr{R} \in \mathbb{R}^{N\times N}$ is symmetric and positive definite and $\mathscr{Q}$ is the self adjoint coercive operator in $\mathscr{L}(\Xi_2)$ [CZ95, Ch. 6]. The cost function (3.43) can be interpreted equivalently as the standard quadratic cost function for linear time-invariant ODE systems. The weighting matrix $\mathscr{R}$ is a design parameter which specifies how many tasks should be moved among the cores to balance the temperature distribution. The second term of the right hand side equals

$$\begin{aligned} < \widehat{\boldsymbol{\xi}}_2(t), \mathscr{Q}\widehat{\boldsymbol{\xi}}_2(t) > =& < \boldsymbol{\xi}_2(t), \mathscr{Q}\boldsymbol{\xi}_2(t) > + 2 < \boldsymbol{\xi}_2(t), \mathscr{Q}\mathscr{C}_2 \sum_{l=1}^{L_d} \int_{t-l\cdot t_d}^{t} e^{\boldsymbol{A}(t-\tau-l\cdot t_d)} \boldsymbol{B}_l \boldsymbol{P}(\tau) d\tau > \\ &+ < \mathscr{C}_2 \sum_{l=1}^{L_d} \int_{t-l\cdot t_d}^{t} e^{\boldsymbol{A}(t-\tau-l\cdot t_d)} \boldsymbol{B}_l \boldsymbol{P}(\tau) d\tau, \mathscr{Q}\mathscr{C}_2 \sum_{l=1}^{L_d} \int_{t-l\cdot t_d}^{t} e^{\boldsymbol{A}(t-\tau-l\cdot t_d)} \boldsymbol{B}_l(\tau) \boldsymbol{P}(\tau) d\tau > . \end{aligned}$$

As all three terms on the right hand side of the above equation are non-negative the real time temperature difference is considered in the given cost function.

In the following, an output feedback controller is applied to manage the power consumption and balance the temperature. The control design is presented in the following theorem.

Theorem 3.1. *Consider the dynamical system (3.40) with the cost function (3.43). If* *there exist a self-adjoint operator* $\boldsymbol{\Psi} \in \mathscr{L}(\mathscr{T}_{\text{a}})$ *and a matrix* $\boldsymbol{F} \in \mathbb{R}^{N\times N}$ *such that*

$$\begin{aligned}&< \boldsymbol{\Psi}\boldsymbol{T}_{\text{a},m}, (\boldsymbol{A} - \widehat{\boldsymbol{B}}\boldsymbol{F}\mathscr{C}_1)\boldsymbol{T}_{\text{a},n} > + < (\boldsymbol{A} - \widehat{\boldsymbol{B}}\boldsymbol{F}\mathscr{C}_1)\boldsymbol{T}_{\text{a},m}, \boldsymbol{\Psi}\boldsymbol{T}_{\text{a},n} > \\ &\quad + < \mathscr{C}_2\boldsymbol{T}_{\text{a},m}, \mathscr{Q}\mathscr{C}_2\boldsymbol{T}_{\text{a},n} > + < (\boldsymbol{F}\mathscr{C}_1\boldsymbol{T}_{\text{a},m}, \mathscr{R}\boldsymbol{F}\mathscr{C}_1\boldsymbol{T}_{\text{a},n} > = 0,\end{aligned} \tag{3.44}$$

where $\boldsymbol{T}_{\text{a},m}$ *and* $\boldsymbol{T}_{\text{a},n} \in \boldsymbol{D}(\boldsymbol{A})$*, where* $\boldsymbol{D}(\boldsymbol{A})$ *indicates the domain of* $\boldsymbol{A}$ *(see A.6), then the controller can be constructed by*

$$\boldsymbol{P}_{\text{c}}(t) = \boldsymbol{F}\mathscr{C}_1\Big(\boldsymbol{T}_{\text{a}}(t) + \sum_{l=1}^{L_d}\int_{t-l\cdot t_{\text{d}}}^{t} e^{\boldsymbol{A}(t-\tau-l\cdot t_{\text{d}})}\boldsymbol{B}_l\boldsymbol{P}(\tau)d\tau\Big), \tag{3.45}$$

which minimizes the cost function (3.43) under the output feedback control policy.

Proof. Define the output feedback controller as

$$\boldsymbol{P}_{\text{c}}(t) = \boldsymbol{F}\mathscr{C}_1\widehat{\boldsymbol{T}}_{\text{a}}(t), \tag{3.46}$$

and rewrite the cost function J as

$$\begin{aligned}J &= \int_0^\infty \Big[< \boldsymbol{P}_{\text{c}}(t), \mathscr{R}\boldsymbol{P}_{\text{c}}(t) > + \int_0^{L_z} < \widehat{\boldsymbol{\xi}}_2(z,t), \mathscr{Q}\widehat{\boldsymbol{\xi}}_2(z,t) > dz\Big] dt \\ &= \int_0^\infty \Big[< \boldsymbol{F}\mathscr{C}_1\widehat{\boldsymbol{T}}_{\text{a}}(t), \mathscr{R}\boldsymbol{F}\mathscr{C}_1\widehat{\boldsymbol{T}}_{\text{a}}(t) > + < \mathscr{C}_2\widehat{\boldsymbol{T}}_{\text{a}}(t), \mathscr{Q}\mathscr{C}_2\widehat{\boldsymbol{T}}_{\text{a}}(t) > dz\Big] dt \\ &= \int_0^\infty < \widehat{\boldsymbol{T}}_{\text{a}}(t), (\mathscr{C}_1^T\boldsymbol{F}^T\mathscr{R}\boldsymbol{F}\mathscr{C}_1 + \mathscr{C}_2^T\mathscr{Q}\mathscr{C}_2)\widehat{\boldsymbol{T}}_{\text{a}}(t) > dt \\ &= \int_0^\infty < \widehat{\boldsymbol{T}}_{\text{a}}(t), \widehat{\mathscr{Q}}\widehat{\boldsymbol{T}}_{\text{a}}(t) > dt.\end{aligned}$$

Defining $\boldsymbol{\Psi}$ as a self adjoint, nonnegative operator, according to the infinite-dimensional quadratic optimal control shown in [CZ95, Ch. 6], one has

$$< \boldsymbol{\Psi}\widehat{\boldsymbol{T}}_{\text{a},m}, (\boldsymbol{A} - \widehat{\boldsymbol{B}}\boldsymbol{F}\mathscr{C}_1)\widehat{\boldsymbol{T}}_{\text{a},n} > + < (\boldsymbol{A} - \widehat{\boldsymbol{B}}\boldsymbol{F}\mathscr{C}_1)\widehat{\boldsymbol{T}}_{\text{a},m}, \boldsymbol{\Psi}\widehat{\boldsymbol{T}}_{\text{a},n} > + < \widehat{\boldsymbol{T}}_{\text{a},m}, \widehat{\mathscr{Q}}\widehat{\boldsymbol{T}}_{\text{a},n} >= 0$$

From (3.27) and (3.40), it is obvious that $\mathscr{T}_{\text{a}}$ and the space of the transformed state vector $\widehat{\mathscr{T}_{\text{a}}}$ share the same orthogonal basement, therefore $\widehat{\boldsymbol{T}}_{\text{a},m}, \widehat{T}_{\text{a},n} \in \mathbf{D}(\boldsymbol{A})$. Then it follows

$$\begin{aligned}&< \boldsymbol{\Psi}\boldsymbol{T}_{\text{a},m}, (\boldsymbol{A} - \widehat{\boldsymbol{B}}\boldsymbol{F}\mathscr{C}_1)\boldsymbol{T}_{\text{a},n} > + < (\boldsymbol{A} - \widehat{\boldsymbol{B}}\boldsymbol{F}\mathscr{C}_1)\boldsymbol{T}_{\text{a},m}, \boldsymbol{\Psi}\boldsymbol{T}_{\text{a},n} > \\ &\quad + < \mathscr{C}_2\boldsymbol{T}_{\text{a},m}, \mathscr{Q}\mathscr{C}_2\boldsymbol{T}_{\text{a},n} > + < (\boldsymbol{F}\mathscr{C}_1\boldsymbol{T}_{\text{a},m}, \mathscr{R}\boldsymbol{F}\mathscr{C}_1\boldsymbol{T}_{\text{a},n} > = 0,\end{aligned} \tag{3.47}$$

which completes the proof. □

The definition of $\boldsymbol{D}(\boldsymbol{A})$ is given in Appendix A.6. The details of solving the Riccati equation (3.44) in Theorem 3.1 are shown in Appendix A.7.

Remark 3.2. *In the controller shown in equation (3.45), the actual measured output* $\boldsymbol{\xi}_1(t) = \mathscr{C}_1 \boldsymbol{T}_a(t)$ *as well as previous power vectors* $\boldsymbol{P}(t - l \cdot t_d)$ *are considered. For real implementation the variables are measured only at the sampling instants. Therefore, the integral in equation (3.45) is then calculated by the sum*

$$\boldsymbol{P}_c(t_k) = \underbrace{\boldsymbol{F}\mathscr{C}_1}_{\hat{\boldsymbol{F}}_0} \boldsymbol{T}_a(t_k) + \sum_{l=1}^{L_d} \underbrace{\boldsymbol{F}\mathscr{C}_1 \int_{-l \cdot t_d}^{0} e^{\boldsymbol{A}(\tau)} d\tau \boldsymbol{B}_l}_{\hat{\boldsymbol{F}}_l} \boldsymbol{P}(t_{k-l}), \qquad (3.48)$$

such that a small overhead can be achieved. The matrices $\hat{\boldsymbol{F}}_l$, $l = \{1, 2, ..., L_d\}$ *can be computed offline.*

3.3 Simulation results

For the simulation, the 8-core IBM Cell Processor, which is described in Section 2.4 is employed. Assume that the frequency range of the processor is from 2 GHz to 4.8 GHz, and the power supply is between 0.9 V and 1.3 V [PBB+05]. Then the thermal behavior is modeled as a group of eight 1D PDEs ((3.17) and (3.18)) with 40 delayed input terms. The parameter $L_d = 40$ is chosen based on the time constant $t_c = 1\,\mathrm{s}$ of the system (3.17) and the variable t_d, see equation (3.20). A suitable choice for t_d is setting it equal to the sampling period, see also (3.48) for realization of the controller in this simulation. Hence, an 8×8 optimal thermal balancing controller matrix can be obtained from Theorem 3.1.

During the simulation, the workload is set as time varying (see the target power in Figure 3.6), and two different tasks are set based on [Lov10, Ch. 4]. One type of task is an urgent task with a deadline, and its frequency cannot be changed while the other type is a non-urgent task where its frequency can be changed online. Under the proposed control policy, the temperature difference among the cores is shown by the 2-Norm of the output vector $\|\boldsymbol{\xi}(t)\|_2$ in Figure 3.4.

The temperature difference under the power based temperature management policy in [MMA+08] is also shown in Figure 3.4. The method proposed in [MMA+08] has the aim to manage the temperature by optimizing the power consumption. A cost function is introduced which takes only the power consumption into account but not the temperature difference among the cores. Therefore, the method proposed in this Chapter can obtain a better performance compared with the power based management technique in [MMA+08].

The comparison of the steady-state temperature distribution between the proposed method and the average task allocation policy under same power consumption is shown is Figure 3.5. The figure shows that the proposed method has a smoother temperature distribution. The original target power $\boldsymbol{P}_t(t)$ and the actual power $\boldsymbol{P}(t) = \boldsymbol{P}_t(t) + \boldsymbol{P}_c(t)$ are shown for the cores 3 and 8 in Figure 3.6. Figure 3.5 shows that core 8 has a lower

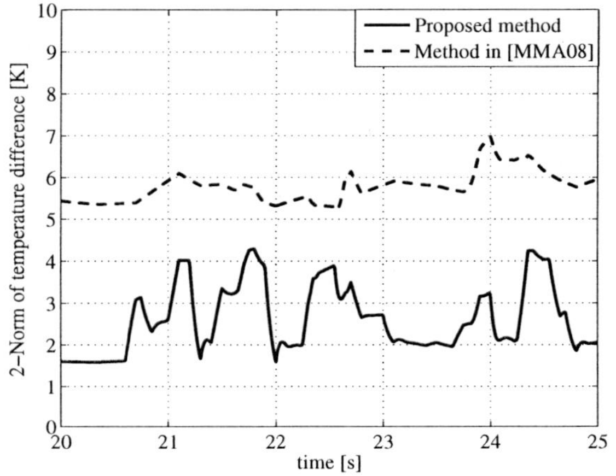

Figure 3.4: Temperature difference among cores

temperature compared with the other cores under a constant workload without a balancing controller while core 3 has a higher temperature. Under the proposed control policy, core 8 works on a higher workload than the target workload. On the other hand core 3 is assigned with a lower workload. Besides, some cores run with fewer assigned tasks than the original target workload while other cores run with more assigned tasks. Therefore, the system still can finish the assigned tasks while the temperature distribution is more balanced.

3.4 Summary

In this chapter, a new model based on a group of 1D PDEs derived from the 3D model developed in Chapter 2 is presented. Thereafter, an optimal control approach is proposed based on the 1D model to balance the temperature and to manage the power consumption among the cores. A Riccati equation approach is introduced to design the output feedback controller. An IBM CELL 8-core processor is employed to demonstrate the effectiveness and efficiency of the proposed control design technique.

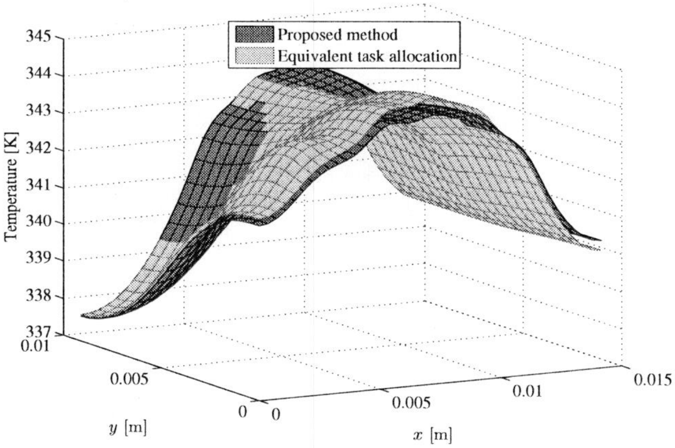

Figure 3.5: Comparison of the temperature distribution

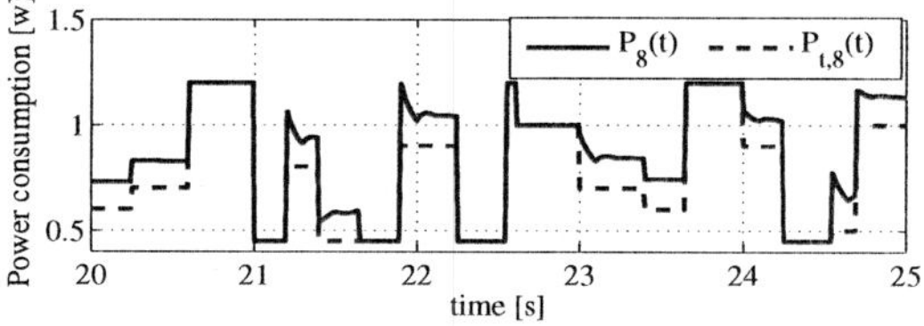

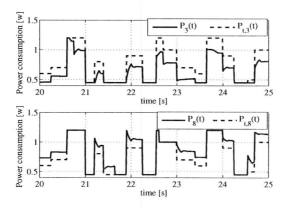

Figure 3.6: Comaprison of the target power and actual power

4 Thermal model of 3D stacked package MCP

As mentioned in Chapter 1, because of the stacking, much more heat will be gathered inside the dies as the power density has a linear relationship with the number of the stacked layers. This is one of the main challenges of the power and thermal management techniques in the 3D chips. Therefore, a cooling system among the layers is a feasible solution to cool the stacked dies inside. A simple structure of a 3D stacked package MCP with micro-channel liquid cooling system (MCLCS) is shown in Figure 4.1 and 4.2.

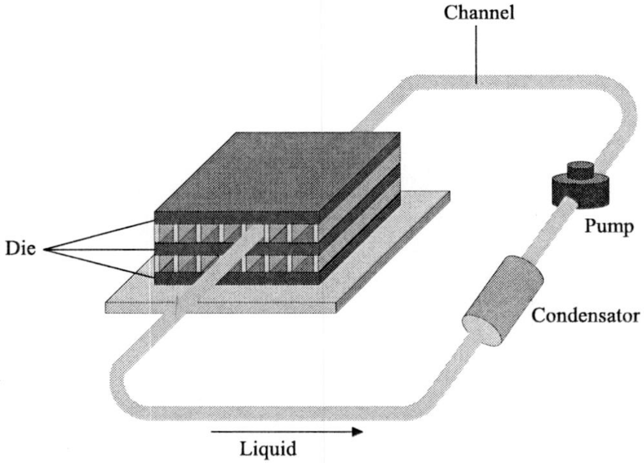

Figure 4.1: 3D package structure

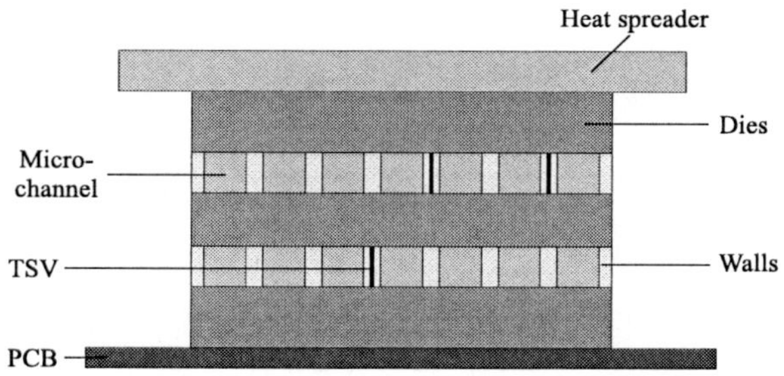

Figure 4.2: The cross-sectional view of the 3D package structure

The integrated micro-channel liquid cooling system (shown in Figure 4.1 and 4.2) developed at EPFL in collaboration with ETHZ and IBM Research has proved to be a feasible solution to cool the 3D ICs [CAAR09, CAR+10, MYL09, QM03, WJ02]. In this chapter, an integrated thermal model, which contains dies, a heat spreader, a micro channel liquid cooling system with a pump, is presented.

4.1 Modeling of the 3D micro-channel liquid cooling system

As mentioned before, the micro-channel liquid cooling system is an effective cooling device for multi-layer MCPs. MCLCS is a circular system driven by a pump. The pump runs to deliver the cool liquid into the stacked layers through a micro-channel. The warmed liquid flows out of the micro-channel, and then is cooled in a condensator. The cool liquid will be brought by the pump again into the micro-channel for another circle. In this thesis, we assume that each system contains one pump for the whole MCLCS and the liquid is delivered through one channel from the pump and then divided into the micro-channels.

4.1.1 Modeling the heat and mass flow in the micro-channels

Compared with the normal size fluid, the micro size fluid has its special characteristics. The micro-channel cooling system causes a phenomenon named thermal wake [ORCP93, MYL09]. For normal size, according to the boundary layer theory, the liquid viscosity only plays a role for a very thin layer near the liquid channel boundary [Sch10, Ch. 11]. Therefore, the upper layer and lower layer do not have cross influence. However, for the micro size flow, the boundary thickness cannot be neglected compared with the channel size. The heat transfer from the upper (lower) layer of the liquid flow upstream may have influence on the heat transfer process of the lower (upper) layer of the liquid flow downstream, which represents the so-called thermal wake, as shown in Figure 4.3.

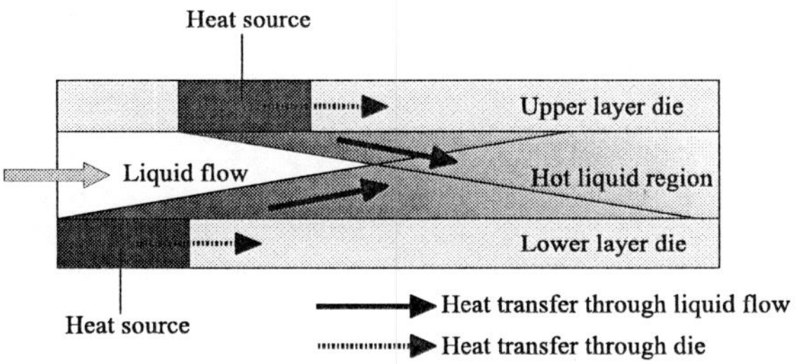

Figure 4.3: Thermal wake

One of the challenges of modeling the thermal behavior in the micro-channel lies in the modeling of the thermal wake phenomena. To model this, it is necessary to obtain the 3D thermal distribution in the micro-channel. Based on the physical facts, the following conditions can be assumed for the liquid fluid.

1. The liquid is an incompressible fluid.
2. The heat radiation is negligible.
3. The flow is a laminar flow.
4. The liquid is a Newtonian fluid.
5. The fluid properties are considered as constant except for the fluid viscosity.

Further explanations are demanded for the assumptions 4 and 5. In assumption 4, the liquid is assumed as Newtonian fluid. Newtonian fluid is a kind of liquid for which the stress versus strain rate curve is linear and passes through the origin [Bat00]. Newtonian fluid can flow under an arbitrarily small external force. Water and air are both Newtonian fluids. In assumption 5, the fluid properties are considered as constant. Normally, water is selected as the cooling liquid. The variations of density, heat conductivity and specific heat capacitance of water are very small and can be neglected. However, the water viscosity strongly depends on the temperature, see Figure 4.4. According to Helmholtz's result, the relationship of water viscosity and temperature can be approached by

$$\mu = \frac{0.001779}{1 + 0.03368(T_\mathrm{f} - 273) + 0.00022099(T_\mathrm{f} - 273)^2}. \tag{4.1}$$

where T_f is the water temperature measured in Kelvin and μ [Pa · s] is the water viscosity

[Hol12].

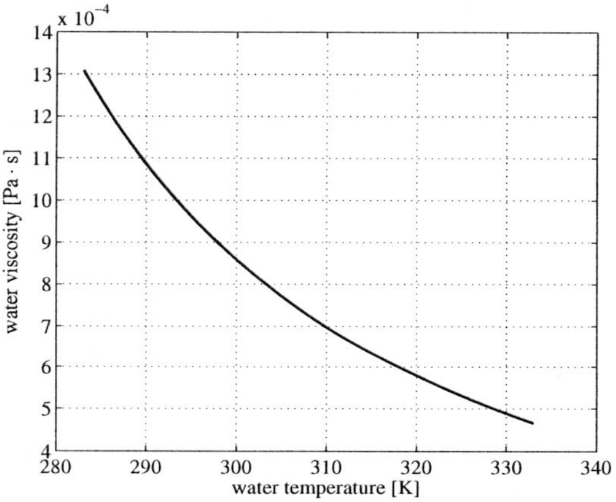

Figure 4.4: Water viscosity along varying temperature

Remark 4.1. *Equation* (4.1) *is a regression result based on the experiment results. The viscosity of water depends both on the temperature and pressure. However, the relation with pressure is very small and can be neglected. Meanwhile, the temperature has a significant influence on the viscosity. Therefore, in equation* (4.1), *only temperature is considered [Hol12, Ch. 11].*

As mentioned in the above assumption, the liquid is incompressible. Therefore, the **conservation of mass** can be described as

$$\frac{\partial u}{\partial x} + \frac{\partial v}{\partial y} + \frac{\partial w}{\partial z} = 0, \tag{4.2}$$

where u, v and w are defined as the velocity components (each with the unit [m/s]) of the liquid flow $\boldsymbol{V}_{\mathrm{s}} = (u \;\; v \;\; w)^T$ in x-, y- and z-direction [Bat00, Sec. 2.3]. The coordinate system is defined in Figure 4.5.

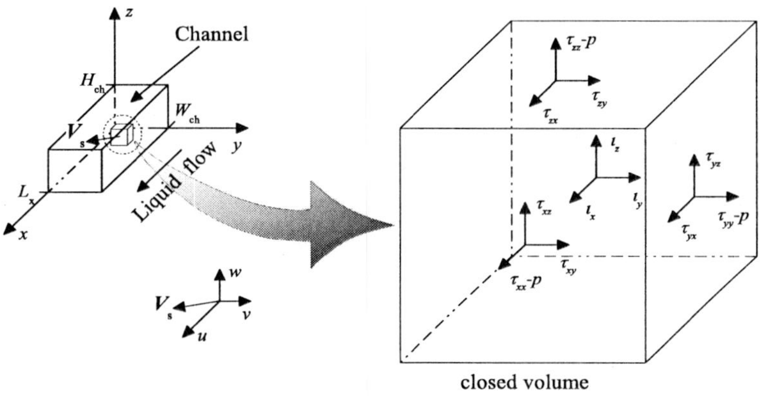

Figure 4.5: The velocity and force analysis of MCLCS

As described in [KIJG05], the channel size is about 10^2 - $10^3\,\mu$m. Therefore, according to the results shown in [QM03, Gad99], the fluid is continuous and the Navier-Stokes equations[3] are still valid to describe the liquid flow behavior. Considering a closed volume in the fluid as shown in Figure 4.5, there are two forces acting on the closed volume, the body force vector $\boldsymbol{\iota} = (\iota_x \ \iota_y \ \iota_z)^T$ with the unit [N/kg] for each component acting on the mass center (see Figure 4.5) and the stress tensor $\boldsymbol{\Gamma} - \boldsymbol{p}$ defined in (4.4) acting on the surfaces [Geb71, Sec. 6.2] (see Figure 4.5 and Appendix B.2 for details). Therefore, the momentum conservation has the form

$$\rho_\mathrm{f}\frac{D\boldsymbol{V}_\mathrm{s}}{Dt} = \nabla\boldsymbol{\Gamma} - \nabla\boldsymbol{p} + \rho_\mathrm{f}\boldsymbol{\iota}, \tag{4.3}$$

where ρ_f [kg/m^3] is the density of the liquid, and $\frac{D\boldsymbol{V}_\mathrm{s}}{Dt}$ is the material derivative with the unit [N/kg] for each component defined by $\frac{D\boldsymbol{V}_\mathrm{s}}{Dt} = \frac{d\boldsymbol{V}_\mathrm{s}}{dt} + \boldsymbol{V}_\mathrm{s}\nabla\boldsymbol{V}_\mathrm{s}$.

For each surface of the closed volume, the stress tensor contains one normal stress $\tau_{ii} - p$ [N/m^2] and two shear stresses τ_{ij} [N/m^2] see Figure 4.5, $\boldsymbol{\Gamma} - \boldsymbol{p}$ has the form

$$\boldsymbol{\Gamma} - \boldsymbol{p} = \begin{bmatrix} \tau_{xx} & \tau_{xy} & \tau_{xz} \\ \tau_{yx} & \tau_{yy} & \tau_{yz} \\ \tau_{zx} & \tau_{zy} & \tau_{zz} \end{bmatrix} - \begin{bmatrix} p & 0 & 0 \\ 0 & p & 0 \\ 0 & 0 & p \end{bmatrix} \tag{4.4}$$

[3]The Navier-Stokes equations are usually applied for Newtonian fluids. In Newtonian fluids the viscous stresses are proportional to the local strain rate, e.g. water [Bat00]. In this thesis water is considered for the cooling system.

where p [N/m^2] is the average pressure in the closed volume. As the liquid is Newtonian fluid, according to the Newtonian law of viscosity, see e.g. [Mor13, Sec. 5.2], one has

$$\tau_{xx} = 2\mu\frac{\partial u}{\partial x}, \tag{4.5a}$$

$$\tau_{yy} = 2\mu\frac{\partial v}{\partial y}, \tag{4.5b}$$

$$\tau_{zz} = 2\mu\frac{\partial w}{\partial z}, \tag{4.5c}$$

and

$$\tau_{xy} = \tau_{yx} = \mu\left(\frac{\partial v}{\partial x} + \frac{\partial u}{\partial y}\right), \tag{4.6a}$$

$$\tau_{xz} = \tau_{zx} = \mu\left(\frac{\partial w}{\partial x} + \frac{\partial u}{\partial z}\right), \tag{4.6b}$$

$$\tau_{yz} = \tau_{yz} = \mu\left(\frac{\partial w}{\partial y} + \frac{\partial v}{\partial z}\right), \tag{4.6c}$$

where μ [Pa $\cdot$ s] is the viscosity of the liquid. Substituting (4.4) - (4.6) into (4.3), one has the momentum conservation equations for the Newtonian fluid

$$\rho_{\mathrm{f}}\left(\frac{\partial u}{\partial t} + u\frac{\partial u}{\partial x} + v\frac{\partial u}{\partial y} + w\frac{\partial u}{\partial z}\right) = \rho_{\mathrm{f}} \iota_x - \frac{\partial p}{\partial x} + \mu\left(\frac{\partial^2 u}{\partial x^2} + \frac{\partial^2 u}{\partial y^2} + \frac{\partial^2 u}{\partial z^2}\right), \tag{4.7a}$$

$$\rho_{\mathrm{f}}\left(\frac{\partial v}{\partial t} + u\frac{\partial v}{\partial x} + v\frac{\partial v}{\partial y} + w\frac{\partial v}{\partial z}\right) = \rho_{\mathrm{f}} \iota_y - \frac{\partial p}{\partial y} + \mu\left(\frac{\partial^2 v}{\partial x^2} + \frac{\partial^2 v}{\partial y^2} + \frac{\partial^2 v}{\partial z^2}\right), \tag{4.7b}$$

$$\rho_{\mathrm{f}}\left(\frac{\partial w}{\partial t} + u\frac{\partial w}{\partial x} + v\frac{\partial w}{\partial y} + w\frac{\partial w}{\partial z}\right) = \rho_{\mathrm{f}} \iota_z - \frac{\partial p}{\partial x} + \mu\left(\frac{\partial^2 w}{\partial x^2} + \frac{\partial^2 w}{\partial y^2} + \frac{\partial^2 w}{\partial z^2}\right). \tag{4.7c}$$

Equation (4.7) describes the liquid flow dynamics which is based on the **conservation of momentum**.

The heat exchange between the liquid in the micro-channel and the die represents the forced heat convection. The thermal behavior can be modeled based on the Fourier's law of heat and energy conservation. Different from the heat conduction in the die, it needs to be considered that the energy crosses though a fixed space volume via the liquid flow.

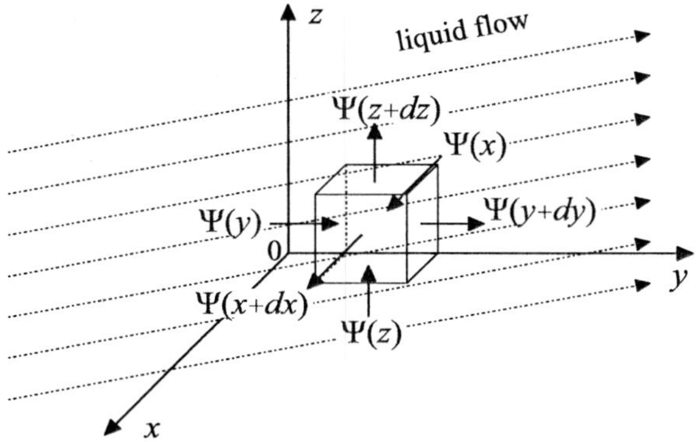

Figure 4.6: Heat transfer in the liquid channel

Selecting a closed small volume in the channel, as shown in Figure 4.6, it is called control volume with fixed coordinates and it is an open system. Based on the first law of thermodynamics [BILD90, Ch. 1 & 6] and [HRW11, Ch. 18], one has

$$\frac{dQ}{dt} = P_\mathrm{f} + m_\mathrm{f,in}\left(H + \frac{1}{2}v_s^2 + gz\right)_\mathrm{in} - m_\mathrm{f,out}\left(H + \frac{1}{2}v_s^2 + gz\right)_\mathrm{out} - P_\mathrm{net}, \tag{4.8}$$

where dQ/dt [J/s] is the internal heat variation per time unit of this volume, P_f [J/s] is the heat flow by conduction heat transfer, that goes from the outside into the closed volume, m_f [kg/s] is the mass flow. The subscript 'in' means that the flow goes inside the area and 'out' means the mass flow goes outside the area. H [J/kg] is the specific enthalpy of the liquid. gz [J/kg] is related to the potential energy (g [N/kg] is gravitational acceleration and z [m] is the position z-direction) and $\frac{1}{2}v_s^2$ [J/kg] to the kinetic energy, where $v_s = \|\boldsymbol{V}_s\|_2$ [m/s]. P_net [J/s] is the work done by the flow per time unit.

In this case we consider a laminar flow, i.e. the liquid velocity v_s and the position z in (4.8) are constant, see also the assumptions given in the beginning of this section. Therefore, the variation of the potential energy and kinetic energy can be neglected. As the flow does not do work the equation (4.8) can be simplified

$$\frac{dQ}{dt} = P_\mathrm{f} + m_\mathrm{f,in}H_\mathrm{in} - m_\mathrm{f,out}H_\mathrm{out}. \tag{4.9}$$

As explained in Section 2.2 the heat conduction rate on a specific point is given by $K\nabla^2 T$, see equation (2.9). As the variable P_f indicates the heat flow by conduction, it is equivalently given by

$$P_\mathrm{f}dt = \int_x^{x+dx}\int_y^{y+dy}\int_z^{z+dz} K_\mathrm{f}\nabla^2 T_\mathrm{f}\,dx\,dy\,dz\,dt \tag{4.10}$$

where K_f [W/(m · K)] is the thermal conductivity and T_f [K] is the temperature of the liquid.

The variation of the interval heat dQ in a closed volume with respect to the variation of the temperature dT_f is given equivalently as in (2.8) by

$$dQ = \int_x^{x+dx} \int_y^{y+dy} \int_z^{z+dz} \sigma_\mathrm{f} \rho_\mathrm{f} \frac{\partial T_\mathrm{f}}{\partial t} dx dy dz dt. \tag{4.11}$$

where σ_f [J/(kg · K)] is the specific heat capacity of the liquid and ρ_f [kg/m^3] is the density of the liquid.

The energy variation in this control volume caused by the mass flow, i.e. $m_{\mathrm{f,in}} H_{\mathrm{in}} - m_{\mathrm{f,out}} H_{\mathrm{out}}$, can be described as the variation of the enthalpy $d\mathcal{H}/dt$ [J/s] caused by the mass flow that goes inside the volume and goes out of the volume. Thereby we distinguish between the mass flow in x-, y- and z-direction. The mass flow in x-direction goes into the closed volume at the point x and goes out of the volume at the point $x+dx$. This is analogously defined for the mass flow in y- and z-direction. Thus, we have

$$m_{\mathrm{f,in}} H_{\mathrm{in}} = \frac{d\mathcal{H}_x}{dt} + \frac{d\mathcal{H}_y}{dt} + \frac{d\mathcal{H}_z}{dt} \tag{4.12a}$$

$$m_{\mathrm{f,out}} H_{\mathrm{out}} = \frac{d\mathcal{H}_{x+dx}}{dt} + \frac{d\mathcal{H}_{y+dy}}{dt} + \frac{d\mathcal{H}_{z+dz}}{dt} \tag{4.12b}$$

For determining $d\mathcal{H}_x/dt$ we need to consider the mass flow in x-direction into the closed volume which is given by

$$m_{x,\mathrm{f,in}} = \int_y^{y+dy} \int_z^{z+dz} u_{\mathrm{f,in}} \rho_\mathrm{f} dy dz \tag{4.13}$$

where $u_{\mathrm{f,in}}$ [m/s] is the liquid velocity in x-direction at the point x. The specific enthalpy H_{in} at the point x is given by

$$H_{\mathrm{in}} = \sigma_\mathrm{f} T_{\mathrm{f,in}} \tag{4.14}$$

such that

$$\frac{d\mathcal{H}_x}{dt} = \int_y^{y+dy} \int_z^{z+dz} \rho_\mathrm{f} \sigma_\mathrm{f} u T_\mathrm{f} dy dz \tag{4.15}$$

setting $u = u_{\mathrm{f,in}}$ and $T_\mathrm{f} = T_{\mathrm{f,in}}$. Analogously we obtain for $d\mathcal{H}_{x+dx}/dt$

$$\frac{d\mathcal{H}_{x+dx}}{dt} = \int_y^{y+dy} \int_z^{z+dz} \rho_\mathrm{f} \sigma_\mathrm{f} \Big(T_\mathrm{f} + \int_x^{x+dx} \frac{\partial T_\mathrm{f}}{\partial x} dx \Big) \Big(u + \int_x^{x+dx} \frac{\partial u}{\partial x} dx \Big) dy dz \tag{4.16}$$

where $(u + \int_x^{x+dx} \frac{\partial u}{\partial x} dx) = u_{\mathrm{out}}$ and $(T_\mathrm{f} + \int_x^{x+dx} \frac{\partial T_\mathrm{f}}{\partial x} dx) = T_{\mathrm{f,out}}$. For determining the energy variation $m_{\mathrm{f,in}} H_{\mathrm{in}} - m_{\mathrm{f,out}} H_{\mathrm{out}}$ we separately determine the energy variation in x-, y- and z-direction first. In x-direction we have

$$\frac{d\mathcal{H}_x}{dt} - \frac{d\mathcal{H}_{x+dx}}{dt} = - \int_x^{x+dx} \int_y^{y+dy} \int_z^{z+dz} \sigma_\mathrm{f} \rho_\mathrm{f} \Big(u \frac{\partial T_\mathrm{f}}{\partial x} + T_\mathrm{f} \frac{\partial u}{\partial x} \Big) dx dy dz, \tag{4.17}$$

The term $\int_x^{x+dx} \frac{\partial u}{\partial x} dx \cdot \int_x^{x+dx} \frac{\partial T_\mathrm{f}}{\partial x} dx$ is neglected in (4.17) as dx is infinitesimal. Following the same derivation we have

$$\frac{d\mathcal{H}_y}{dt} - \frac{d\mathcal{H}_{y+dy}}{dt} = -\int_x^{x+dx}\int_y^{y+dy}\int_z^{z+dz} \sigma_\mathrm{f}\rho_\mathrm{f}\left(v\frac{\partial T_\mathrm{f}}{\partial y} + T_\mathrm{f}\frac{\partial v}{\partial y}\right) dxdydz \tag{4.18}$$

in y-direction and

$$\frac{d\mathcal{H}_z}{dt} - \frac{d\mathcal{H}_{z+dz}}{dt} = -\int_x^{x+dx}\int_y^{y+dy}\int_z^{z+dz} \sigma_\mathrm{f}\rho_\mathrm{f}\left(w\frac{\partial T_\mathrm{f}}{\partial z} + T_\mathrm{f}\frac{\partial w}{\partial z}\right) dxdydz \tag{4.19}$$

Substituting (4.17), (4.18) and (4.19) into (4.9) taking (4.12) into account and further substituting (4.10) and (4.11) into (4.9), one has the **conservation of energy** equation

$$\sigma_\mathrm{f}\rho_\mathrm{f}\left(\frac{\partial T_\mathrm{f}}{\partial t} + u\frac{\partial T_\mathrm{f}}{\partial x} + v\frac{\partial T_\mathrm{f}}{\partial y} + w\frac{\partial T_\mathrm{f}}{\partial z}\right) = K_\mathrm{f}\nabla^2 T_\mathrm{f}. \tag{4.20}$$

Thereby, also the conservation of mass is used, see equation (4.2). The first term of the left hand side is the temperature change with time. The following three terms are the heat variation due to the mass flow. The right hand side is the energy change based on the heat conduction. In summary, the equations (4.2), (4.7) and (4.20) are the model of the liquid flow dynamic and thermal behavior. The boundary conditions of these equations are given in the following. For the hydraulic boundaries except the channel inlet and outlet, the velocity is always zero [QM02, QM03, Sch10, Ch. 11], i.e. for a general liquid flow in x-direction, see Figure 4.5, we have

$$u(x,y,0,t) = u(x,y,H_\mathrm{ch},t) = 0, \quad u(x,0,z,t) = u(x,W_\mathrm{ch},z,t) = 0 \tag{4.21a}$$
$$v(x,y,0,t) = v(x,y,H_\mathrm{ch},t) = 0, \quad v(x,0,z,t) = v(x,W_\mathrm{ch},z,t) = 0 \tag{4.21b}$$
$$w(x,y,0,t) = w(x,y,H_\mathrm{ch},t) = 0, \quad w(x,0,z,t) = w(x,W_\mathrm{ch},z,t) = 0 \tag{4.21c}$$

where H_ch [m] is the height of the channel and W_ch [m] is the width of the channel. For the inlet, it has

$$u(0,y,z,t) = u_\mathrm{in}, \quad v(0,y,z,t) = 0, \quad w(0,y,z,t) = 0. \tag{4.22}$$

Supposing the flow is fully developed at the channel outlet, see Figure 4.7, the boundary conditions are

$$\frac{\partial u(L_x,y,z,t)}{\partial x} = 0, \quad \frac{\partial v(L_x,y,z,t)}{\partial x} = 0, \quad \frac{\partial w(L_x,y,z,t)}{\partial x} = 0 \tag{4.23}$$

where L_x is the length of the channel. For the thermal boundary conditions of the four walls, it is compliant with Newton's laws of cooling, i.e.

$$K\frac{\partial T_\mathrm{f}(x,y,0,t)}{\partial z} = -h_\mathrm{f}\big(T_\mathrm{d}(x,y,0,t) - T_\mathrm{f}(x,y,0,t)\big) \tag{4.24a}$$

$$K\frac{\partial T_\mathrm{f}(x,y,H_\mathrm{ch},t)}{\partial z} = h_\mathrm{f}\big(T_\mathrm{d}(x,y,H_\mathrm{ch},t) - T_\mathrm{f}(x,y,H_\mathrm{ch},t)\big) \tag{4.24b}$$

$$K\frac{\partial T_\mathrm{f}(x,0,z,t)}{\partial y} = -h_\mathrm{f}\big(T_\mathrm{w}(x,0,z,t) - T_\mathrm{f}(x,0,z,t)\big) \tag{4.24c}$$

$$K\frac{\partial T_\mathrm{f}(x,W_\mathrm{ch},z,t)}{\partial y} = h_\mathrm{f}\big(T_\mathrm{w}(x,W_\mathrm{ch},z,t) - T_\mathrm{f}(x,W_\mathrm{ch},z,t)\big) \tag{4.24d}$$

where $T_\mathrm{d}(x,y,z,t)$ is the temperature of the die, $T_\mathrm{w}(x,y,z,t)$ is the temperature of the wall between the channels and $T_\mathrm{f}(x,y,z,t)$ is the temperature of the liquid. At the channel inlet, the liquid temperature is equal to a fixed constant, i.e. $T(0,y,z,t) = T_\mathrm{in}$. Suppose the thermal is fully developed at the channel outlet, we have

$$\frac{\partial^2 T_\mathrm{f}(L_x,y,z,t)}{\partial x^2} = 0. \tag{4.25}$$

To get the thermal and dynamical behavior of the liquid flow in the micro-channel, the governing differential equations (4.2), (4.7) and (4.20) with the boundary conditions (4.21a) to (4.25) can be discretized along the x-, y- and z-direction and the Semi-Implicit Method for Pressure Linked Equations (SIMPLE) algorithm [Pat80] can be used to solve the 3D pressure, velocity and temperature distribution. SIMPLE is an iterative algorithm based on 'Guess-Correct', see Appendix C for the details.

Remark 4.2. *The boundary conditions shown in* (4.23) *and* (4.25) *may not be satisfied, if the channel is very short. However, as the local heat convection coefficient* h_f*, the boundary layer, and the length of the liquid fluid and thermal developing region are not influenced by the length of the channel, these boundary conditions can still be used to investigate the fluid and thermal dynamic behavior of the liquid in the micro-channel, by assuming the channel is long enough that the liquid can reach the fluid developed region and the thermal developed region.*

The heat convection coefficient h_f is a time-varying coefficient, and depends on the liquid viscosity, the Reynolds number, the Prandtl number and the micro-channel position. Further, the fluid needs some length to develop the velocity profile and thermal profile after entering the micro-channel, which are called fluid entrance region and thermal entrance region [BILD90, Ch. 9]. In the fluid entrance region, the fluid flow is developing, and the fluid boundary layer is developing and growing thicker. When the boundary layer is as thick as the radius, the boundary meets at a point and the fluid status becomes stable. From this point the fully developed region starts. The fluid flow characteristic is shown in Figure 4.7, where the length of the arrows represents the liquid velocity in x-direction.

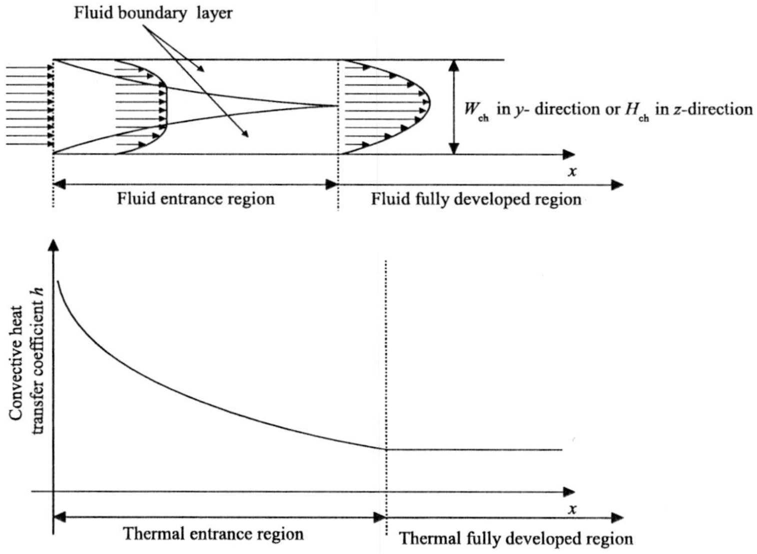

Figure 4.7: Fluid and thermal entrance region

The thermal entrance region is a region where the thermal boundary layer is developing. In the thermal entrance region, the heat transfer coefficient h_{f} is mutative, but in the thermal developed region, h_{f} is constant. The fluid and thermal entrance region length are affected by the channel size, the Reynolds number (Re) and the Prandtl number (Pr). The fluid entrance length is equal to the thermal entrance length only if $Pr = 1$. In this thesis we consider water as the cooling liquid for which the Prandtl number is $Pr = 7$. Thus, the fluid entrance length and the thermal entrance length are not equal. The heat convection coefficient h_{f} can be obtained from

$$h_{\mathrm{f}}(x) = \frac{Nu(x)K_{\mathrm{f}}}{D_{\mathrm{h}}}, \tag{4.26}$$

where Nu is the Nusselt number, D_{h} is the channel hydraulic diameter [HKG99]. According to [HKG99] it can be calculated by $D_{\mathrm{h}} = 2H_{\mathrm{ch}}W_{\mathrm{ch}}/(H_{\mathrm{ch}} + W_{\mathrm{ch}})$. In order to determine the Nusselt number practical experiments are required which is discussed in [QM03]. The concepts of Reynolds number, Prandtl number and Nusselt number are given in Appendix B.1.

Remark 4.3. *According to [SB87, KC05] the length of the fluid entrance region* L_{f} [m] *can be approached by*

$$L_{\mathrm{f}} \approx 0.05 Re D_{\mathrm{h}} \tag{4.27}$$

while the length of the thermal entrance region L_t [m] can be approached by

$$L_\mathrm{t} \approx 0.05 Re Pr D_\mathrm{h}. \tag{4.28}$$

Those approximations are valid for laminar fluids.

Remark 4.4. *As here the fluid is assumed as laminar, the convection heat transfer coefficient is increasing with liquid velocity. According to [KMS13, Sec. 6.3], enhancing the fluid velocity has a positive influence on the heat convection process from the die to the liquid.*

Assume that the chip is installed in the horizontal plane. Figure 4.8 shows the fluid velocity and temperature characteristics in a micro-channel. Figure 4.8(a) shows a draft of the distribution of the fluid velocity at $x = x_1$ and $y = \frac{W_\mathrm{ch}}{2}$ indicated by the dashed line in the channel assuming there is a fluid entrance flow. Additionally the distribution of the fluid velocity is given for a fully developed flow at $x = x_2$ and $y = \frac{W_\mathrm{ch}}{2}$ [KZYW09, QM02]. The fluid velocity in y- and z- direction is very small compared with the velocity u in x-direction, and can be neglected. Therefore, the average value of fluid velocity in x-direction is approximately $u = u_\mathrm{in}$ [KZYW09, QM02]. In normal-size channels, the temperature has a sudden change in the boundary layer, but in the middle of the channel, the temperature changes barely [KMS13], which is shown in Figure 4.8(c). However, in a micro-channel the heat spreads in the whole channel as shown in Figure 4.8(b), the temperature distribution on a micro-channel cross has similar values, such that the temperature of the water can be considered equally distributed, and it only changes in x-direction [HKG99, QM02, KIJG05].

Therefore, for the following derivation the conservation of energy (4.20) can be simplified to

$$\sigma_\mathrm{f}\rho_\mathrm{f}S_\mathrm{c}\left(\frac{\partial T_\mathrm{f}(x,t)}{\partial t} + u_\mathrm{in}\frac{\partial T_\mathrm{f}(x,t)}{\partial x}\right) = S_\mathrm{c}K_\mathrm{f}\nabla^2 T_\mathrm{f}(x,t). \tag{4.29}$$

where $S_\mathrm{c} = W_\mathrm{ch}H_\mathrm{ch}$ is the cross-section area of the channel. Equation (4.29) with the unit $[\mathrm{J}/(\mathrm{s}\cdot\mathrm{m}]$ describes the heat variation dynamics with respect to time and the position x. However equation (4.29) does not consider the heat exchange of the channel with the dies and the wall along the channel boundaries. In order to include the heat exchange along the boundaries the conditions (4.24) are employed. The heat increase due to the heat exchange with the bottom die, i.e. $z = 0$, at the given point x in x-direction is given by

$$\int_0^{W_\mathrm{ch}} -K\frac{\partial T_\mathrm{f}(x,t)}{\partial z}dy = \int_0^{W_\mathrm{ch}} h_\mathrm{f}\big(T_\mathrm{d,down}(x,t) - T_\mathrm{f}(x,t)\big)dy = -h_\mathrm{f}W_\mathrm{ch}\big(T_\mathrm{f}(x,t) - T_\mathrm{d,down}(x,t)\big). \tag{4.30}$$

according to (4.24a). The temperature of the liquid $T_\mathrm{f}(x,t)$ is constant with respect to y and z as explained in the above paragraph. Along the boundary line at one cross section of the micro-channel the temperature of the die and the wall is assumed to be constant with respect to y and z, see Figure 4.9. In the later derivation die and the liquid channel

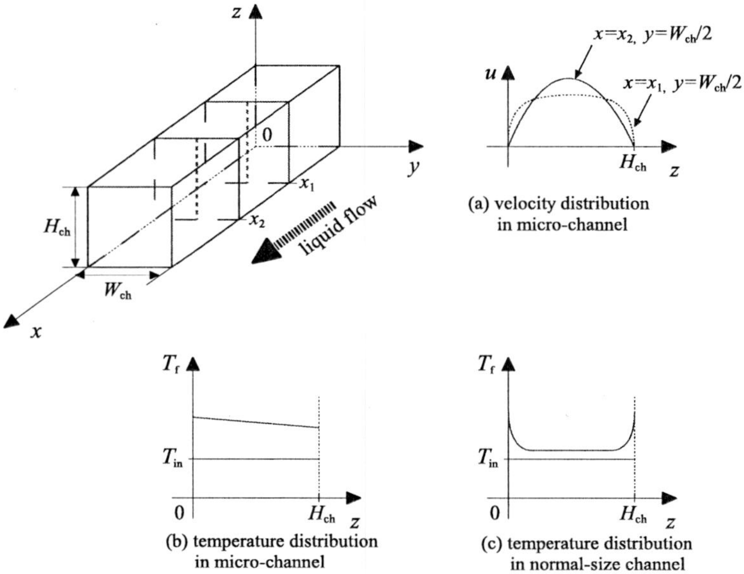

Figure 4.8: Fluid velocity and temperature distribution in a micro-channel

is divided into blocks such that this assumption is still valid. Based on (4.24b) at the level $z = H_{\mathrm{ch}}$ the heat increase due to the heat exchange with upper die is given by

$$\int_0^{W_{\mathrm{ch}}} K \frac{\partial T_{\mathrm{f}}(x,t)}{\partial z} dy = -h_{\mathrm{f}} W_{\mathrm{ch}} \big(T_{\mathrm{f}}(x,t) - T_{\mathrm{d,up}}(x,t)\big) \tag{4.31}$$

In an analogous way the heat exchange with the walls can be determined based on (4.24c)-(4.24d). In the following it needs to be considered that the convective heat transfer coefficient h_{f} varies with respect to x, u_{in} and T_{f} as well as the temperature of the boundary blocks, i.e. T_{d} and T_{w}.

Thus, the heat variation dynamics with respect to time and the position x are given by

$$\begin{aligned} \sigma_{\mathrm{f}}\rho_{\mathrm{f}}S_c\left(\frac{\partial T_{\mathrm{f}}}{\partial t} + u_{\mathrm{in}}\frac{\partial T_{\mathrm{f}}}{\partial x}\right) &= S_c K_{\mathrm{f}} \frac{\partial^2 T_{\mathrm{f}}}{\partial x^2} \\ - h_{\mathrm{f}}\Big(W_{\mathrm{ch}}(T_{\mathrm{f}} - T_{\mathrm{d,up}}) + W_{\mathrm{ch}}(T_{\mathrm{f}} - T_{\mathrm{d,down}}) &+ H_{\mathrm{ch}}(T_{\mathrm{f}} - T_{\mathrm{w1}}) + H_{\mathrm{ch}}(T - T_{\mathrm{w2}})\Big) \end{aligned} \tag{4.32}$$

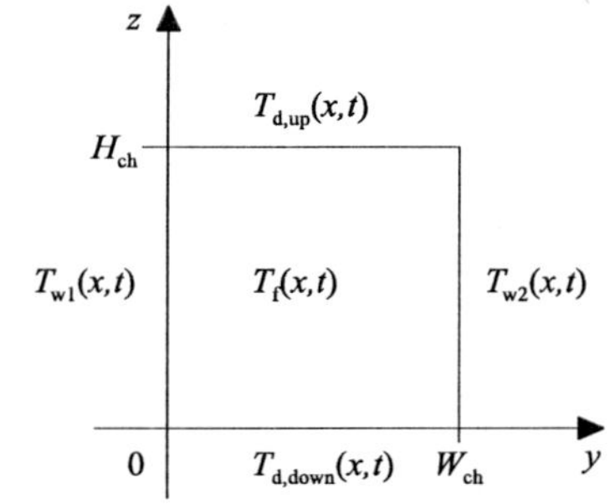

Figure 4.9: Cross section of the channel at a position x

Considering the physical fact that the heat transfer in the liquid is mainly caused by the heat convection and the fluid flow, the heat conduction term of the liquid in the right hand side can be omitted. The energy conservation equation can then be written as

$$h_\mathrm{f}\Big(W_\mathrm{ch}(T_\mathrm{f}-T_\mathrm{d,up})+W_\mathrm{ch}(T_\mathrm{f}-T_\mathrm{d,down})+H_\mathrm{ch}(T_\mathrm{f}-T_\mathrm{w1})+H_\mathrm{ch}(T-T_\mathrm{w2})\Big)$$
$$+\sigma_\mathrm{f}\rho_\mathrm{f}S_\mathrm{c}\left(\frac{\partial T_\mathrm{f}}{\partial t}+u_\mathrm{in}\frac{\partial T_\mathrm{f}}{\partial x}\right)=0 \qquad (4.33)$$

In equation (4.33) all temperatures are assumed constant with respect to y and z and vary with respect to x and t. In order to derive an ODE system a discretization in x-direction is conducted for both the micro-channel and the dies, see Figure 4.10. Thereby the channel and the die is divided to several blocks. The size of the blocks is not uniform, as the convective heat transfer coefficient varies in the channel significantly, especially

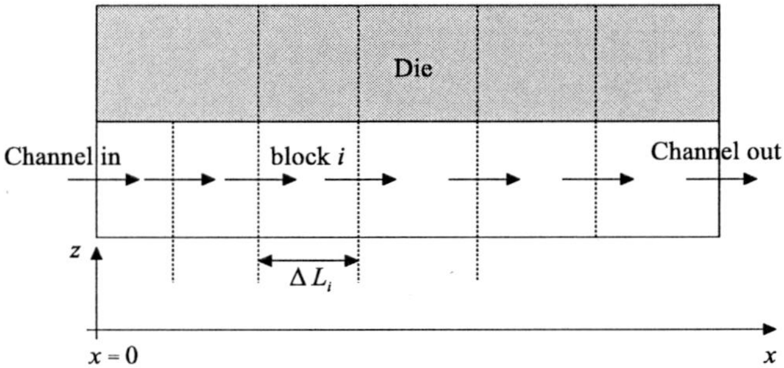

Figure 4.10: Discretization of the micro-channel and the die

in the heat entrance region. To get a more precise model, in the heat entrance region, the channel should be divided in a smaller size. Additionally the die blocks need to be considered while discretizing the micro-channel such that above/under one micro-channel block the temperature of the die is constant with respect to x, see Figure 4.10. The discretization of the dies is discussed in Section 4.2.

For each discrete block, the energy variation is caused by the heat convection from the channel walls to the channel and the fluid flow from the upstream as the heat conduction term is ignored. Assuming the temperature of the block i is $T_{\mathrm{f},i}(t)$, and each grid length is ΔL_i, see Figure 4.10, one has

$$\begin{aligned} dT_{\mathrm{f},i}(t) =& \frac{u_{\mathrm{in}}(t)T_{\mathrm{f},i-1}(t) - u_{\mathrm{in}}(t)T_{\mathrm{f},i}(t)}{\Delta L_i}dt + \frac{\Delta L_i}{V_{\mathrm{f},i}\rho_{\mathrm{f}}\sigma_{\mathrm{f}}}\Big(h_{\mathrm{f},i}(t)W_{\mathrm{ch}}(T_{\mathrm{d},j_{\mathrm{up}}}(t) - T_{\mathrm{f},i}(t)) \\ &+ h_{\mathrm{f},i}(t)W_{\mathrm{ch}}(T_{\mathrm{d},j_{\mathrm{down}}}(t) - T_{\mathrm{f},i}(t)) + h_{\mathrm{f},i}(t)H_{\mathrm{ch}}(T_{\mathrm{w},j_1}(t) - T_{\mathrm{f},i}(t)) \\ &+ h_{\mathrm{f},i}(t)H_{\mathrm{ch}}(T_{\mathrm{w},j_2}(t) - T_{\mathrm{f},i}(t))\Big)dt \end{aligned} \tag{4.34}$$

where $V_{\mathrm{f},i}$ is the volume of block i, $T_{\mathrm{d},j_{\mathrm{up}}}$, $T_{\mathrm{d},j_{\mathrm{down}}}$, T_{w,j_1} and T_{w,j_2} are the temperatures of the die and wall blocks which are corresponding to channel block i respectively. In equation (4.34) the convective heat transfer coefficient $h_{\mathrm{f},i}(t)$ is constant with respect to the position in each block i but varies with respect to time due to its dependency on the temperatures and the liquid speed.

Therefore, the system model is given by the differential equation

$$\begin{aligned} \dot{T}_{\mathrm{f},i}(t) =& -\left(\frac{u_{\mathrm{in}}(t)}{\Delta L_i} + 2\frac{\Delta L_i W_{\mathrm{ch}} h_{\mathrm{f},i}(t)}{V_{\mathrm{f},i}\rho_{\mathrm{f}}\sigma_{\mathrm{f}}} + 2\frac{\Delta L_i H_{\mathrm{ch}} h_{\mathrm{f},i}(t)}{V_{\mathrm{f},i}\rho_{\mathrm{f}}\sigma_{\mathrm{f}}}\right)T_{\mathrm{f},i}(t) \\ &+ \frac{u_{\mathrm{in}}(t)}{\Delta L_i}T_{\mathrm{f},i-1}(t) + \frac{\Delta L_i W_{\mathrm{ch}} h_{\mathrm{f},i}(t)}{V_{\mathrm{f},i}\rho_{\mathrm{f}}\sigma_{\mathrm{f}}}T_{\mathrm{d},j_{\mathrm{up}}}(t) + \frac{\Delta L_i W_{\mathrm{ch}} h_{\mathrm{f},i}(t)}{V_{\mathrm{f},i}\rho_{\mathrm{f}}\sigma_{\mathrm{f}}}T_{\mathrm{d},j_{\mathrm{down}}}(t) \\ &+ \frac{\Delta L_i H_{\mathrm{ch}} h_{\mathrm{f},i}(t)}{V_{\mathrm{f},i}\rho_{\mathrm{f}}\sigma_{\mathrm{f}}}T_{\mathrm{w},j_1}(t) + \frac{\Delta L_i H_{\mathrm{ch}} h_{\mathrm{f},i}(t)}{V_{\mathrm{f},i}\rho_{\mathrm{f}}\sigma_{\mathrm{f}}}T_{\mathrm{w},j_2}(t) \end{aligned} \tag{4.35}$$

which is the space discrete MCLCS model.

4.1.2 Modeling of the pump and the liquid flow velocity

As shown in Figure 4.1, the liquid cooling system is driven by an pump. We assume that the Laing 12V DC pump is a suitable choice for the liquid cooling system of the 3D package systems [Lai10]. The cool liquid from the pump flows to each micro-channel, and then goes into the chip under the same flow rate [CAR+10]. Assume the micro-channels have the same size. According to the pump curve shown in [Lai10], the pump flow rate V_{pump} [m^3/s] and the pump output power P_{pump} [W] can be approximated by the linear equation $V_{\mathrm{pump}} = K_{\mathrm{pump}} \cdot P_{\mathrm{pump}}$ with the constant K_{pump} [m^3/J] which can be obtained by the pump curve. According to the mass conservation law, the flow rate of the pump output and the micro-channel has the relationship

$$S_{\mathrm{pump}} u_{\mathrm{pump}} = N_{\mathrm{c}} S_{\mathrm{c}} u_{\mathrm{in}} \tag{4.36}$$

where S_{pump} [m^2] is the cross-sectional area of the pump output channel, u_{pump} [m/s] is the velocity of pump output fluid, N_{c} is the number of channels and $S_{\mathrm{c}} = W_{\mathrm{ch}} \cdot H_{\mathrm{ch}}$ is the cross-sectional area of the micro-channel.

4.2 R-C heat transfer model of the dies

As described in [Kre00], the thermal transfer and the electrical phenomena are dual, see Table 4.1. The heat flow can be considered as the 'current' while the temperature difference is considered as the 'voltage'. Therefore, a heat flow, that goes through a thermal resistance R, leads to a 'voltage'. Meanwhile, in order to describe the setting time before the temperature reaching a steady state in the thermal transfer phenomena, a thermal capacitance C is also introduced. By dividing the die in blocks, each block can be described as an R-C model while the whole system can be described as an R-C network, see Figure 4.11. The blocks may have different volumes, and suppose a block i has the size $L_{xi} \times L_{yi} \times L_{zi}$. The relative thermal resistance and thermal capacitance can be calculated as

$$R_{\mathrm{w}i} = R_{\mathrm{e}i} = \frac{0.5 L_{xi}}{K L_{yi} L_{zi}}, \tag{4.37a}$$

$$R_{\mathrm{n}i} = R_{\mathrm{s}i} = \frac{0.5 L_{yi}}{K L_{xi} L_{zi}}, \tag{4.37b}$$

$$R_{\mathrm{u}i} = R_{\mathrm{d}i} = \frac{0.5 L_{zi}}{K L_{xi} L_{yi}}, \tag{4.37c}$$

$$C_{\mathrm{t}i} = \rho \sigma L_{xi} L_{yi} L_{zi}. \tag{4.37d}$$

Therefore, the whole die can be divided into a finite number of blocks, and each block is approached by a single thermal R-C model. By connecting these R-C models together, an R-C network is achieved to describe thermal behavior of the whole die.

Thermal quantity	unit	Electrical quantity	unit
P, power	W	I, Current	A
T, Temperature difference	K	V, Voltage	V
R, Thermal resistance	K/W	R, Electrical resistance	$\Omega = \mathrm{V/A}$
C, Thermal mass, capacitance	J/K	C, Electrical capacitance	$\mathrm{F} = \mathrm{A/V}$
RC, Thermal RC constant	s	RC, Electrical RC constant	s

Table 4.1: Duality between thermal and electrical quantities [SSH+03b]

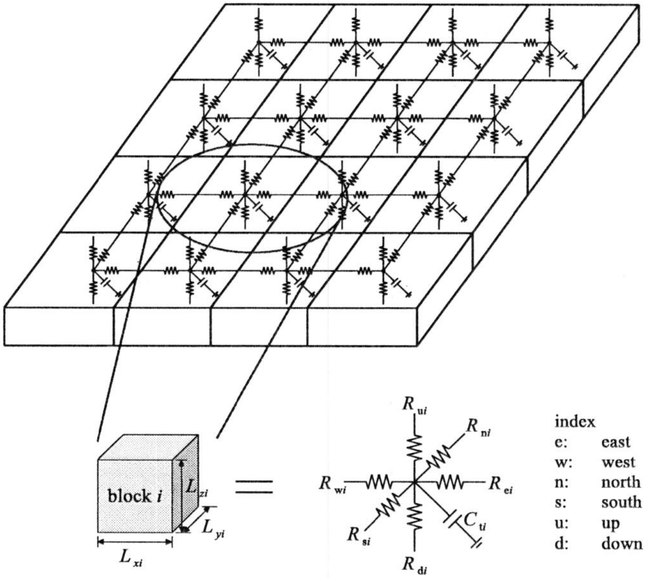

Figure 4.11: Modeling of the die as an R-C network

An example of the R-C network is shown in Figure 4.12. The die is divided to two blocks and the channel divided to 3 blocks. The mathematical description is given by

$$C_{\mathrm{t}1}\frac{dT_{\mathrm{d},1}}{dt}+\frac{T_{\mathrm{d},1}-T_{\mathrm{d},2}}{R_{\mathrm{e}1}+R_{\mathrm{w}2}}+\frac{T_{\mathrm{d},1}-T_{\mathrm{f},1}}{R_{\mathrm{d}1}+R_{\mathrm{f}1}}+\frac{T_{\mathrm{d},1}-T_{\mathrm{f},2}}{R_{\mathrm{d}1}+R_{\mathrm{f}2}}+\frac{T_{\mathrm{d},1}-T_{\infty}}{R_{\mathrm{u}1}+R_{\mathrm{a}1}}=P_1 \tag{4.38}$$

$$C_{\mathrm{t}2}\frac{dT_{\mathrm{d},2}}{dt}+\frac{T_{\mathrm{d},2}-T_{\mathrm{d},1}}{R_{\mathrm{e}1}+R_{\mathrm{w}2}}+\frac{T_{\mathrm{d},2}-T_{\mathrm{f},3}}{R_{\mathrm{d}2}+R_{\mathrm{f}3}}+\frac{T_{\mathrm{d},2}-T_{\infty}}{R_{\mathrm{u}2}+R_{\mathrm{a}2}}=P_2 \tag{4.39}$$

In equation (4.38), the first term $C_{\mathrm{t}1}\frac{dT_{\mathrm{d},1}}{dt}$ describes the temperature variation, the second term $\frac{T_{\mathrm{d},1}-T_{\mathrm{d},2}}{R_{\mathrm{e}1}+R_{\mathrm{w}2}}$ describes the heat flow between die block 1 and die block 2, $\frac{T_{\mathrm{d},1}-T_{\mathrm{f},1}}{R_{\mathrm{d}1}+R_{\mathrm{f}1}}$ describes the heat flow between the die block 1 and the channel block 1, $\frac{T_{\mathrm{d},1}-T_{\mathrm{f},2}}{R_{\mathrm{d}1}+R_{\mathrm{f}2}}$ is the heat flow between the die block 1 and the channel block 2, $\frac{T_{\mathrm{d},1}-T_{\infty}}{R_{\mathrm{u}1}+R_{\mathrm{a}1}}$ is the heat flow between the die block 1 and the environment while P_1 is the internal heat source. $R_{\mathrm{f}1}$ and $R_{\mathrm{f}2}$ are time varying and can be gotten as [KB95]

$$R_{\mathrm{f}k}=\frac{1}{S_{\mathrm{d}k}h_{\mathrm{f},k}(t)}. \tag{4.40}$$

In equation (4.39), P_2 is the internal heat source and $R_{\mathrm{f}3}$ can be also determined based on (4.40). Equation (4.38) and (4.39) can be written as

$$
\begin{aligned}
\dot{T}_{\mathrm{d},1} =& \frac{1}{C_{\mathrm{t1}}}\left(\frac{1}{R_{\mathrm{e1}}+R_{\mathrm{w2}}}+\frac{1}{R_{\mathrm{d1}}+R_{\mathrm{f1}}}+\frac{1}{R_{\mathrm{d1}}+R_{\mathrm{f2}}}+\frac{1}{R_{\mathrm{u1}}+R_{\mathrm{a1}}}\right)T_{\mathrm{d},1} \\
&+\frac{1}{C_{\mathrm{t1}}(R_{\mathrm{e1}}+R_{\mathrm{w2}})}T_{\mathrm{d},2}+\frac{1}{C_{\mathrm{t1}}(R_{\mathrm{d1}}+R_{\mathrm{f1}})}T_{\mathrm{f},1}+\frac{1}{C_{\mathrm{t1}}(R_{\mathrm{d1}}+R_{\mathrm{f2}})}T_{\mathrm{f},2} \\
&+\frac{1}{C_{\mathrm{t1}}(R_{\mathrm{u1}}+R_{\mathrm{a1}})}T_{\infty}+\frac{1}{C_{\mathrm{t1}}}P_1, \qquad (4.41\mathrm{a}) \\
\dot{T}_{\mathrm{d},2} =& -\frac{1}{C_{\mathrm{t2}}}\left(\frac{1}{R_{\mathrm{e1}}+R_{\mathrm{w2}}}+\frac{1}{R_{\mathrm{d2}}+R_{\mathrm{f3}}}+\frac{1}{R_{\mathrm{u2}}+R_{\mathrm{a2}}}\right)T_{\mathrm{d},2}+\frac{1}{C_{\mathrm{t2}}(R_{\mathrm{e1}}+R_{\mathrm{w2}})}T_{\mathrm{d},1} \\
&+\frac{1}{C_{\mathrm{t2}}(R_{\mathrm{d2}}+R_{\mathrm{f3}})}T_{\mathrm{f},3}+\frac{1}{C_{\mathrm{t2}}(R_{\mathrm{u2}}+R_{\mathrm{a2}})}T_{\infty}+\frac{1}{C_{\mathrm{t2}}}P_2. \qquad (4.41\mathrm{b})
\end{aligned}
$$

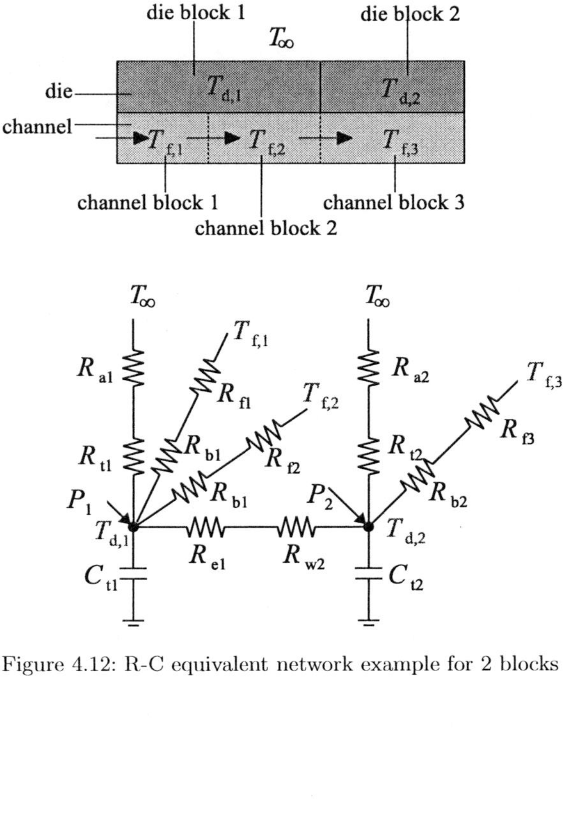

Figure 4.12: R-C equivalent network example for 2 blocks

In summary, for each block, the thermal behavior is described by the equation

$$\dot{T}_{\mathrm{d},i} = -\frac{1}{C_{\mathrm{t}i}R_i}T_{\mathrm{d},i} + \frac{1}{C_{\mathrm{t}i}}\sum_{j=1}^{\mathscr{J}}\frac{1}{R_j}T_{\mathrm{d},j} + \frac{1}{C_{\mathrm{t}i}}\sum_{k=1}^{\mathscr{K}}\frac{1}{R_k}T_{\mathrm{f},k} + \frac{1}{C_{\mathrm{t}i}R_{i\infty}}T_\infty + \frac{1}{C_{\mathrm{t}i}}P_i \tag{4.42}$$

where

$$\frac{1}{R_i} = \sum_{j=1}^{\mathscr{J}}\frac{1}{R_j} + \sum_{k=1}^{\mathscr{K}}\frac{1}{R_k} + \frac{1}{R_{i\infty}}, \tag{4.43}$$

$$R_j = \begin{cases} R_{\mathrm{e}i} + R_{\mathrm{w}j} & \text{block } j \text{ is in the east of block } i, \\ R_{\mathrm{w}i} + R_{\mathrm{e}j} & \text{block } j \text{ is in the west of block } i, \\ R_{\mathrm{s}i} + R_{\mathrm{n}j} & \text{block } j \text{ is in the south of block } i, \\ R_{\mathrm{n}i} + R_{\mathrm{s}j} & \text{block } j \text{ is in the north of block } i, \end{cases} \tag{4.44}$$

and $\mathscr{J} \in \mathbb{N}$ is the number of neighbors of block i. Besides,

$$R_k = R_{\mathrm{u}i} + R_{\mathrm{f}k} \tag{4.45}$$

and $\mathscr{K} \in \mathbb{N}$ is the number of channel blocks on the top and bottom of the ith die block. The heat transfer coefficient to the ambient air is

$$R_{i\infty} = R_{\mathrm{a}i} + R_{\mathrm{u}i} \tag{4.46}$$

and $R_{ai} = \infty$, if upper and lower of block i sides both convect the heat to the micro-channel.

4.3 Model integration

By integrating the model of the dies and the channels, thermal behavior model of the whole system can be achieved. The number of channels is normally 10 - 100. If we consider each divided block in every channel as an independent state, the number of the states of the whole system will be a very large, which will result in a complex online control algorithm, and not suitable for the real time control. Therefore, the channels with similar thermal behavior, i.e. the channels with the same upper die block and lower die block can be grouped as shown in Figure 4.13. Basically a trade-off between model complexity and preciseness of the model needs to be found. Meanwhile, the wall influence is ignored as compared with the channel, the heat transfer via the wall is very small such that the influence on the die and channel temperature can be neglected.

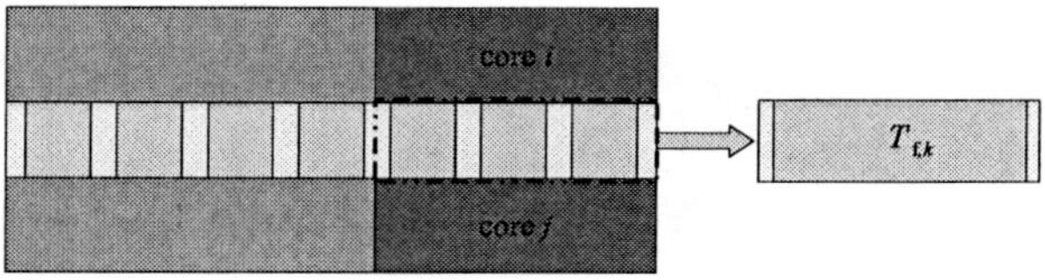

Figure 4.13: The grouped channels

Therefore, define the system state vector as

$$\boldsymbol{T} = \begin{bmatrix} \boldsymbol{T}_{\mathrm{d}} \\ \boldsymbol{T}_{\mathrm{f}} \end{bmatrix} \tag{4.47}$$

where

$$\boldsymbol{T}_{\mathrm{d}} = [T_{\mathrm{d},1}, T_{\mathrm{d},2}, ..., T_{\mathrm{d},M_1}]^T$$

is the state vector of the blocks in the die and M_1 is the number of blocks of the dies.

$$\boldsymbol{T}_{\mathrm{f}} = [T_{\mathrm{f},1}, T_{\mathrm{f},2}, ..., T_{\mathrm{f},M_2}]^T$$

is the state vector of the micro-channel and M_2 is the number of blocks of the channels. Thereafter, based on (4.35) and (4.42), the system describing the whole network is given by

$$\dot{\boldsymbol{T}}(t) = \boldsymbol{A}\big(\boldsymbol{T}(t), u_{\mathrm{in}}(t)\big)\boldsymbol{T}(t) + \boldsymbol{B}_1\boldsymbol{P}(t) + \boldsymbol{B}_2\boldsymbol{d}(t) + \boldsymbol{B}_3T_\infty + \boldsymbol{B}_4T_{\mathrm{in}} \tag{4.48}$$

where the vector $\boldsymbol{P}(t) = [P_1(t), P_2(t), ..., P_N(t)]^T$ is considered as the power consumption of the cores and N is the number of cores. The power can be controlled by the supply voltage and frequency of the cores see Section 2.1. T_∞ is the environment temperature, while T_{in} is the liquid inlet temperature. The time varying system matrix is $\boldsymbol{A} \in \mathbb{R}^{(M1+M2)\times(M1+M2)}$, the control input matrix is $\boldsymbol{B}_1 \in \mathbb{R}^{(M1+M2)\times N}$, the disturbance input matrix is $\boldsymbol{B}_2 \in \mathbb{R}^{(M1+M2)\times(M1+M2)}$, $\boldsymbol{B}_3 \in \mathbb{R}^{(M1+M2)}$ and $\boldsymbol{B}_4 \in \mathbb{R}^{(M1+M2)}$ are the input matrices of the constant temperature of the ambient temperature T_∞ and the liquid inlet temperature T_{in}. $\boldsymbol{d}(t)$ is considered as an unknown input vector, which contains the power consumption of non-core blocks, for instance the L2 cache and I/O controller in Figure 2.6. $\boldsymbol{d}(t)$ is an unknown but bounded vector, which satisfies

$$\boldsymbol{d}(t) \geq 0 \quad \text{and} \quad \int_0^\infty \boldsymbol{d}^T(t)\boldsymbol{d}(t) < \infty. \tag{4.49}$$

4.4 Example of the thermal modeling

The IBM CELL processor shown in Chapter 2 is employed as the simulation object to verify the model. Suppose the system has 3 layers, the first and second layers contain 8 cores, which is the same as the IBM CELL processor. The L2 cache, the power processor element and the I/O are in the third layers as shown in Figure 4.14. The numbers in Figure 4.14 and 4.15 indicate the state vector index for the simulation.

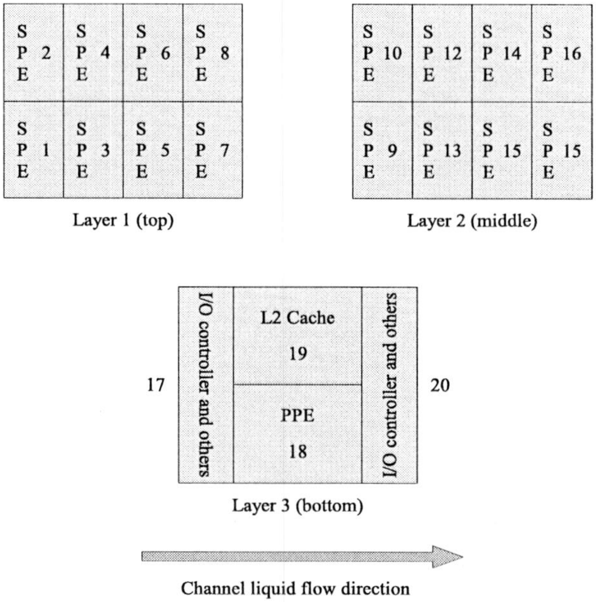

Figure 4.14: Die layout plan in the simulation

number of layers	3
die size (same of three layers)	10 mm × 9.1 mm
micro-channel width (W_{ch})	700 μm
micro-channel height (H_{ch})	300 μm
wall width	100 μm
overall number of micro-channels	20
number of blocks in the first and second layers	8
number of blocks in the third layer	4

Table 4.2: Parameters of the 3D IC example test structure

The channel size and other parameters are shown in the Table 4.2, while the parameters relative to the micro-channel cooling system are designed based on [MYL09]. The three layers are divided in 20 blocks as shown in Figure 4.14. Each micro-channel cooling layer contains 10 micro-channels. We consider them as two groups, see Figure 4.15. Each micro-channel cooling layer is divided in 10 blocks. Therefore, a state vector with 40 states is defined.

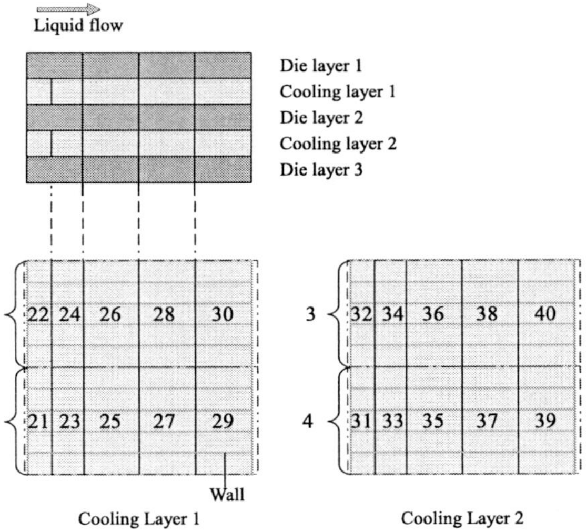

Figure 4.15: Die layout plan in the simulation

The Reynolds number, the Nusselt number and heat convection coefficient are influenced by the water temperature as shown in Figure 4.16 - 4.18. The Nusselt number and the convective heat transfer coefficient additionally vary with respect to the position x in the cooling micro-channel and the liquid velocity. Figure 4.17 - 4.18 show the Nusselt number and the convective heat transfer coefficient under a constant liquid velocity $u_{\mathrm{in}} = 0.5\,\mathrm{m/s}$.

The results show that the temperature affects the Nusselt number and the convective heat transfer coefficient. This is explained by the influence of the liquid temperature on the water viscosity as mentioned before. Further the water viscosity influences the Nusselt number and the convective heat transfer coefficient. The convective heat transfer coefficient is larger in the thermal entrance region, i.e. for small x, than in the thermal developed region. When the liquid reaches the thermal developed region, the heat convection coefficient can be considered as constant. For the micro-channel of the MCLCS the length of the fluid entrance region is $L_{\mathrm{f}} \approx 10\,\mathrm{mm}$ and the length of the thermal entrance region $L_{\mathrm{t}} \approx 70\,\mathrm{mm}$ for a liquid with the temperature $T_{\mathrm{f}} \approx 298\,\mathrm{K}$. As the length of the micro-channel ($L_x = 10\,\mathrm{mm}$) is approximately equal to the fluid entrance region and smaller than the thermal entrance region the regions are not fully developed.

Therefore, the heat convection coefficient must be considered as varying with respect to the position.

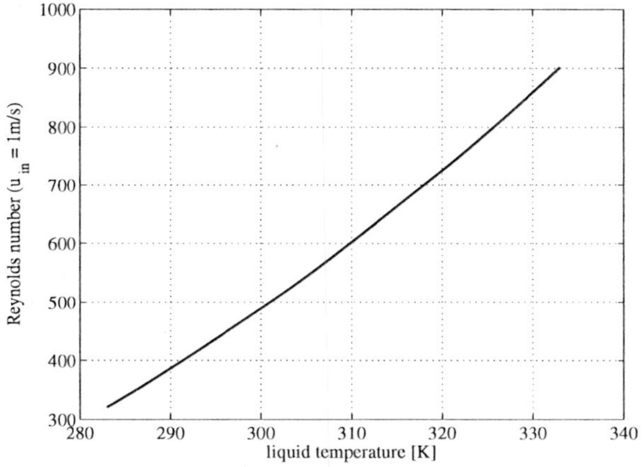

Figure 4.16: Reynolds number along varying liquid temperature

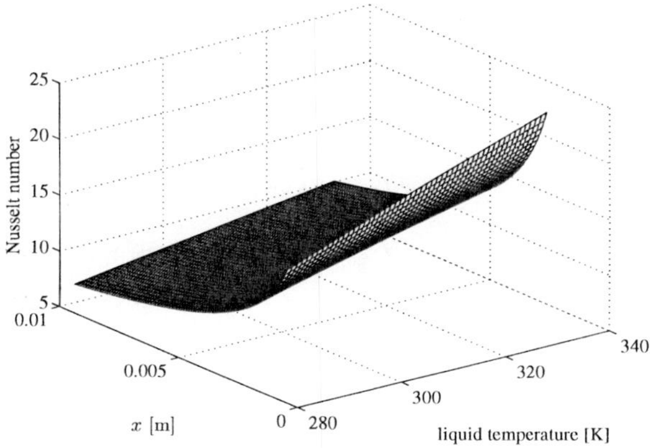

Figure 4.17: Nusselt number along varying liquid temperature and position

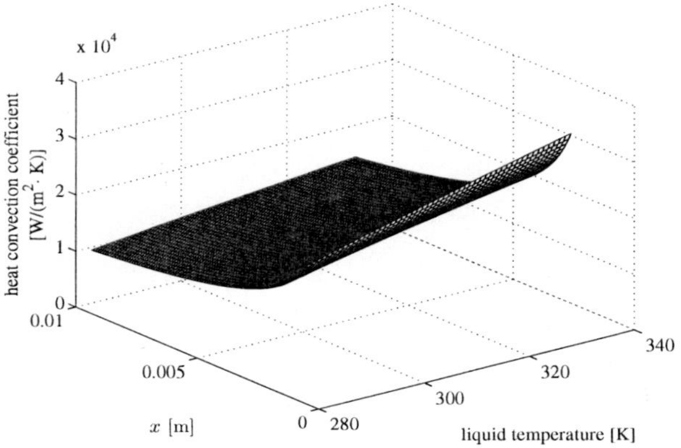

Figure 4.18: Convective heat transfer coefficient along varying liquid temperature and position

Figure 4.19 - 4.21 show the variation of the Nusselt number and the convective heat transfer coefficient with respect to the position and the liquid velocity under a constant temperature of 298 K. A quicker liquid velocity indicates better heat transfer ability.

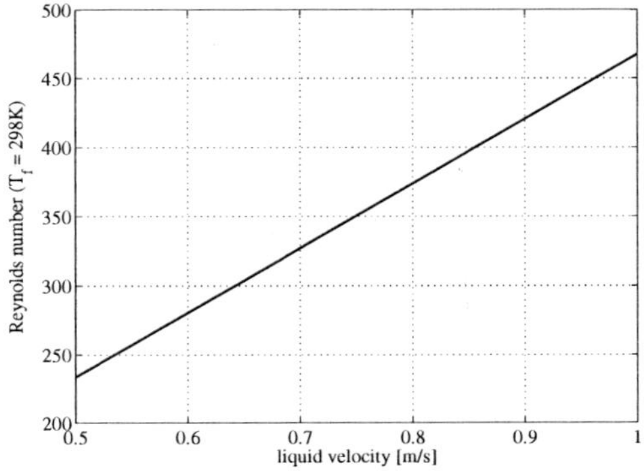

Figure 4.19: Reynolds number along varying liquid velocity

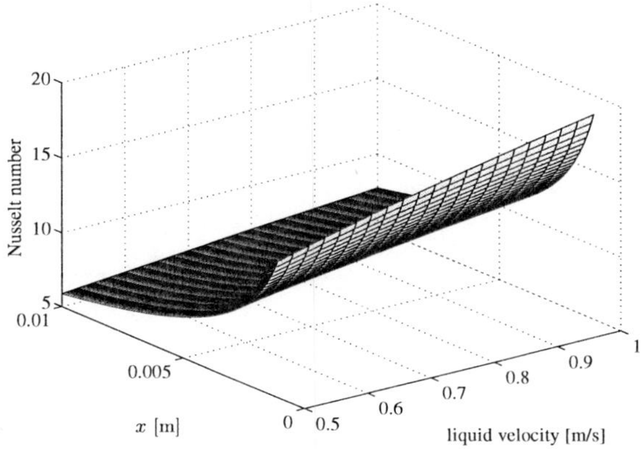

Figure 4.20: Nusselt number along varying liquid velocity and position

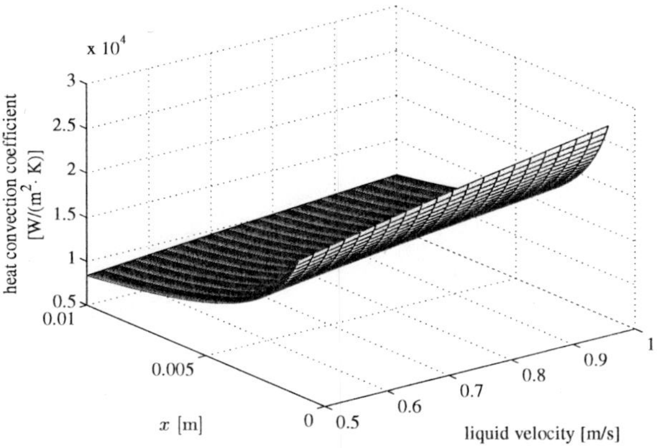

Figure 4.21: Convective heat transfer coefficient along varying liquid velocity and position

Remark 4.5. *As explained in Section 4.1.1 the channel is discretized into blocks. In that model the heat transfer coefficient is assumed constant within one block. For achieving an accurate simulation the blocks need to be assigned appropriately, see the discussion in Section 4.1.1.*

To obtain the system response, a step power input is set. Each block of the dies has the same power density, which means that all the cores run with the same voltage and frequency. Besides, the input of the blocks in the bottom layer is the double of the one in the first and second layer. The other disturbance inputs are also set as constant power consumptions. The environment temperature is set 298 K, the input water temperature is assumed 293 K, and the cooling channel water velocity is $u_{\text{in}} = 0.5\,\text{m/s}$. Under this power consumption, the cores with symmetrical position, for instance, core 1 and core 2, should have the same thermal behavior.

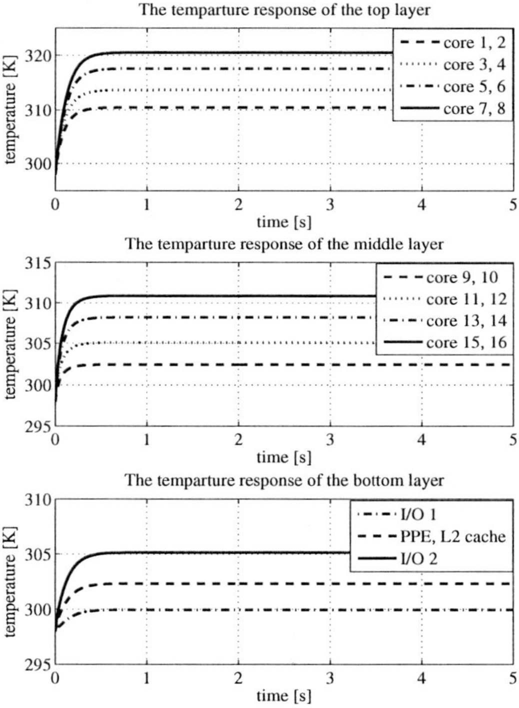

Figure 4.22: Step input thermal behavior of the dies under $u_{\text{in}} = 0.5\,\text{m/s}$

As shown in Figure 4.22, under the same power consumption, the lowest temperature is in the channel inlet side of the middle layer die and the highest temperature is in the channel outlet side of the bottom layer, because the water forced heat convection has a better heat exchanging ability than the air heat convection. As the water forced heat convection coefficient is very large (10^4) compared with the air heat convection coefficient (10^2 - 10^3), the main heat escapes in the liquid cooling system.

The thermal behavior of the cooling channel inside is shown in Figure 4.23. From this figure we see that the liquid blocks 21 and 22 have the lowest temperature while the blocks 39 and 40 have the highest temperature. This is because the blocks 21 and 22 are next to the inlet, blocks 39 and 40 are next to the outlet. Further more the top die has a better heat escaping condition than the bottom die.

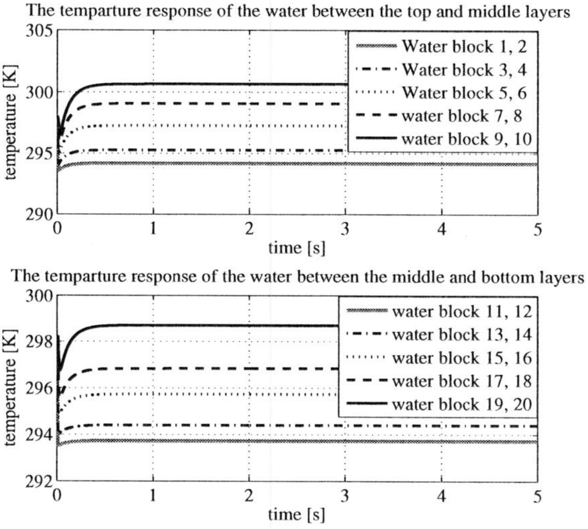

Figure 4.23: Step input thermal behavior of the fluid under $u_{\text{in}} = 0.5\,\text{m/s}$

In the following simulation results are shown for varying tasks, i.e. a time varying target power, which is defined equivalently for each core, see Figure 4.24.

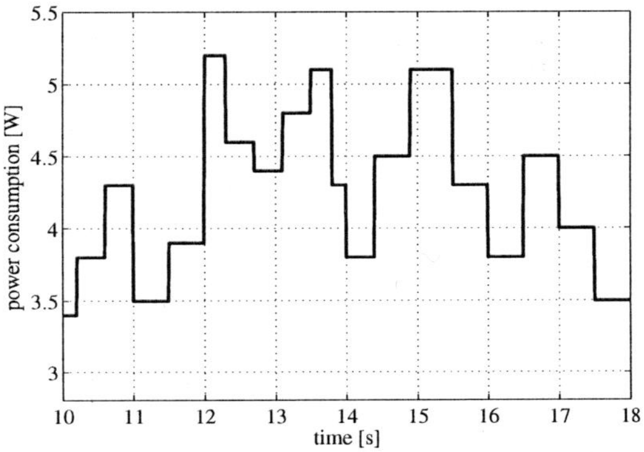

Figure 4.24: Time-varying target power set equal for each core

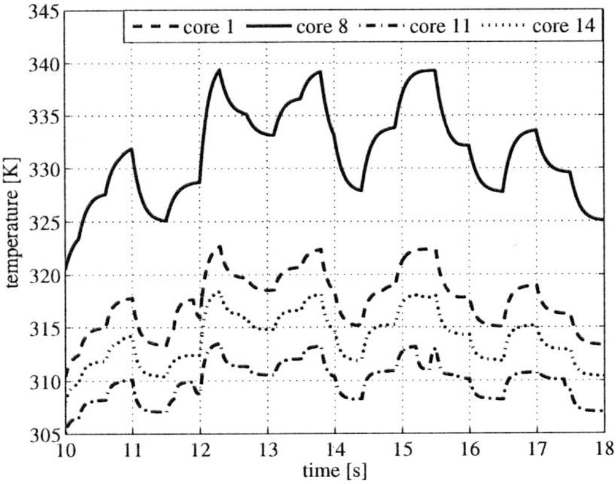

Figure 4.25: Real time thermal behavior of the dies under $u_{\text{in}} = 0.5\,\text{m/s}$

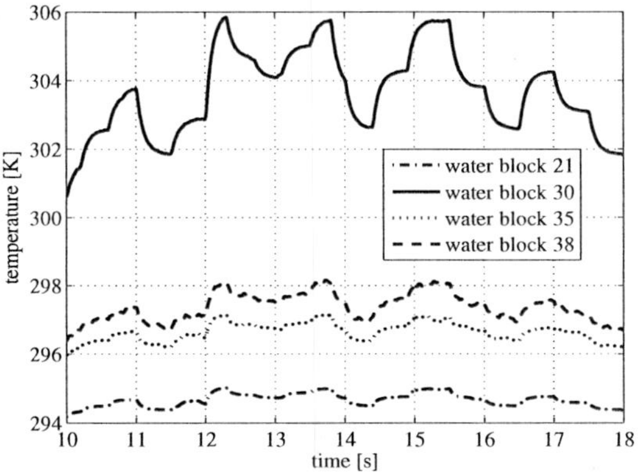

Figure 4.26: Real time thermal behavior of the fluid under $u_{\text{in}} = 0.5\,\text{m/s}$

In Figure 4.25 and 4.26, the real time thermal response with varying tasks is presented. The operation condition of the multi-core processor is defined equivalent to Section 3.3 with two different kind of tasks. Under this condition, core 8 has the highest temperature, and the cores in the second die have a lower temperature, which shows an equivalent trend of the distribution of the heat as in Figure 4.22.

Figure 4.27 shows the thermal behavior of the dies under the channel liquid velocity $u_{\text{in}} = 1\,\text{m/s}$, and the power consumption and operation condition are set the same as in the Figure 4.22. Under this velocity, the water temperature is lower as shown in Figure 4.28 and the convective heat transfer coefficients are higher as shown in Figure 4.21. Therefore, the whole dies temperature is lower than under $u_{\text{in}} = 0.5\,\text{m/s}$. The real time dynamical thermal behavior of the dies and fluid with varying tasks (with the target power given in Figure 4.24) is shown in Figure 4.29 and Figure 4.30, and the power consumption and operation condition are set the same as in Figure 4.25 and Figure 4.26. The temperature shown in these figures is also lower than in the thermal response with $u_{\text{in}} = 0.5\,\text{m/s}$. Therefore, adjusting the fluid velocity is also a possible method to control the temperature.

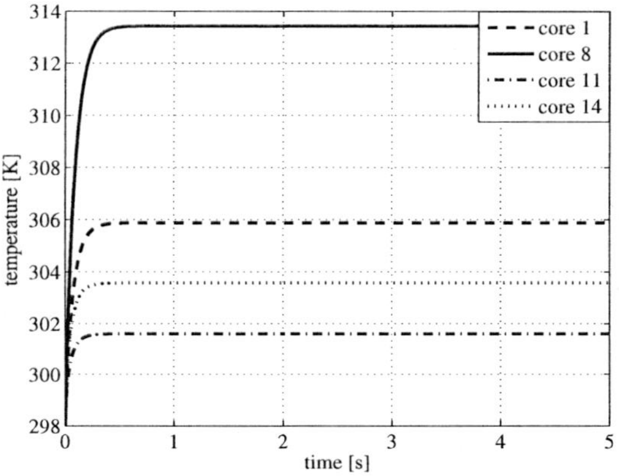

Figure 4.27: Step input thermal behavior of the fluid under $u_{\text{in}} = 1.0\,\text{m/s}$

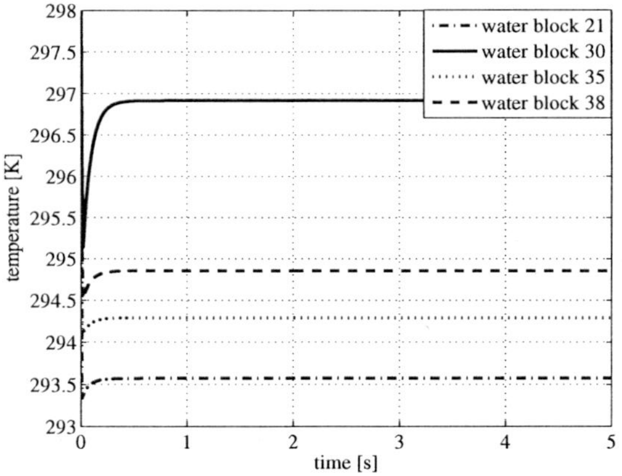

Figure 4.28: Step input thermal behavior of the fluid under $u_{\text{in}} = 1.0\,\text{m/s}$

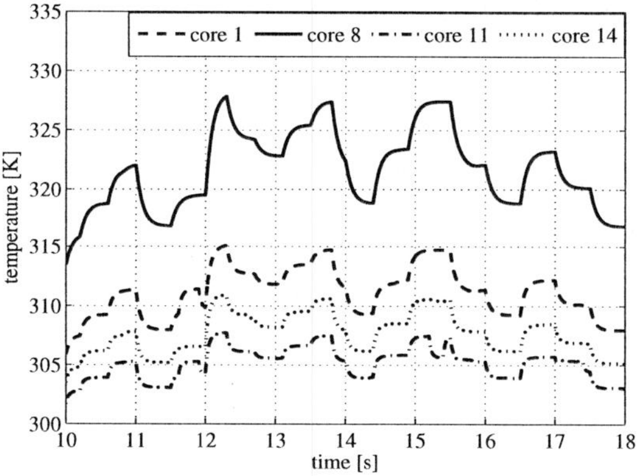

Figure 4.29: Real time thermal behavior of the dies under $u_{\text{in}} = 1.0\,\text{m/s}$

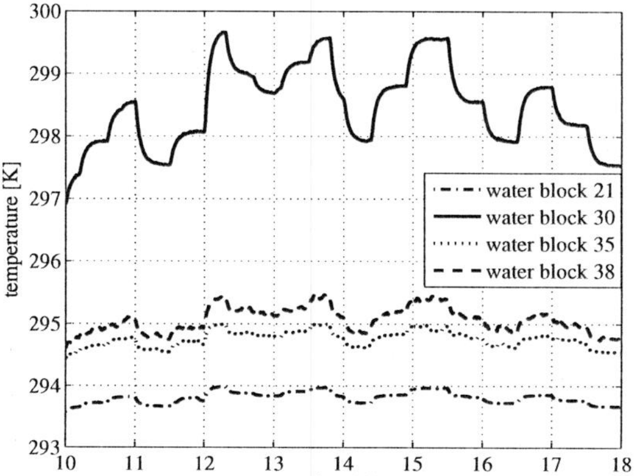

Figure 4.30: Real time thermal behavior of the fluid under $u_{\text{in}} = 1.0\,\text{m/s}$

4.5 Summary

This chapter presents the thermal model of the 3D stacked package MCPs. The model of the dies is described as an R-C network. The cooling channels are discretized based on the micro-channel thermal and liquid dynamic feature. The integrated thermal model of the dies, the cooling channels, the channel walls (containing TSV) and the pump is developed. The simulation results illustrate that the proposed model is consistent with the reality and can be employed for the control design which is discussed in the next chapter.

5 Thermal and power balancing/management policy design for 3D MCP

The control objective for the system described in equation (4.48) is to manage and balance the temperature and power consumption among the different cores. There are two possible ways to manage the temperature and power consumption. One is to adjust the cores supply voltage and frequency via DVFS technology. Another is to adjust the fluid flow velocity, whereas the fluid flow velocity is equal in all micro-channels. As discussed in Chapter 4, in a certain fluid velocity region, the convective heat transfer coefficient increases with the velocity as shown in Figure 4.21. Besides, the temperature difference between the liquid and the dies is larger as shown in the simulation results in Chapter 4, which has positive impact on the forced heat convection.

However, as shown in the system model (4.48), the liquid velocity has a nonlinear influence, which leads to a high complexity for both the control design and the online voltage and frequency adjustment. Therefore, in order to reduce the complexity, the controller is designed in two parts. The cooling liquid input velocity is adjusted under a simple logic algorithm, which is based on the real time work tasks and the highest temperature of the die. Thus, the adjustment of liquid velocity focuses on keeping the temperature of the dies in a temperature range appropriate for the chip's operation reliability and lifespan. The cooling liquid velocity is then not contained in the online optimal control policy, which focuses on balancing the temperature and the power consumption among the cores.

Based on the velocity adjustment policy, the system can be described as a set of independent models corresponding to each possible liquid velocity, i.e. the system is modeled as a switched linear system and the switching law is given by the logic algorithm. The control structure is shown in Figure 5.1. In this figure, $\boldsymbol{\xi}_1$ is the measurement output vector. Based on $\boldsymbol{\xi}_1$, all states can be observed by the H_∞ observer, and $\widehat{\boldsymbol{T}}$ is the estimated state vector. Then based on the estimated state vector $\widehat{\boldsymbol{T}}$ and the target power $\boldsymbol{P}_\mathrm{t}$ defined in (3.39), a switched controller is applied to select the input liquid velocity and the controller, where δ is the switching law. The switching law δ decides which controller and which observer gain matrix to select. The controller output is the adjusting power vector $\boldsymbol{P}_\mathrm{c}$, which can be used to vary the core supply voltage and frequency as described in Chapter 2.

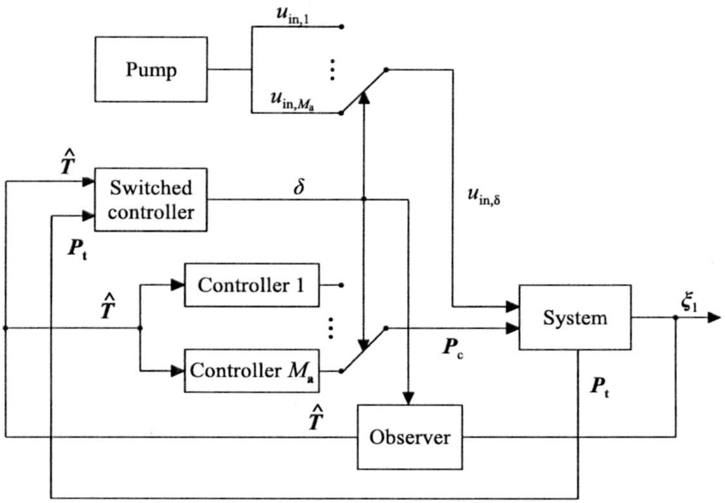

Figure 5.1: The Control structure

5.1 Policy for adjusting the liquid velocity

The basic idea of this algorithm is that the velocity should be increased if the temperature of the dies is too high or more tasks are assigned to the cores, while the velocity should be reduced if the temperature of the dies is too low or fewer tasks are assigned to the cores. Further, the algorithm needs to be designed in a way such that effects as oscillation of the temperature and sticking of the temperature to an undesired temperature region are avoided. The fluid velocity $u_{\text{in}}(t)$ can be set to a finite number of constants

$$u_{\text{in}}(t) \in \mathbb{U}_{\text{in}}, \quad \mathbb{U}_{\text{in}} = \{u_{\text{in},1}, u_{\text{in},2}, ..., u_{\text{in},M_\text{a}}\} \tag{5.1}$$

where $u_{\text{in},1}, u_{\text{in},2}, ..., u_{\text{in},M_\text{a}}$ are constants, and $u_{\text{in},1} < u_{\text{in},2} < ... < u_{\text{in},i} < ... < u_{\text{in},M_\text{a}}$. $u_{\text{in},i}$ can be decided by the pump characteristic and the possible chip load. The difference between each adjacent $u_{\text{in},i}$ can be set as constant but not necessarily.

Define an average target input power

$$P_\text{t}(t) = \frac{1}{N}\sum_{i=1}^{N} P_{\text{t},i}(t). \tag{5.2}$$

where $P_{\text{t},i}(t)$ is the target power of core i. Corresponding to the velocity set (5.1), the target input power $P_t(t)$ is also divided in M_a sets as

$$\begin{aligned}\mathbb{P}_{t,1} &= \{P_t(t)|P_t(t) < P_{s,1}\}\\ \mathbb{P}_{t,i} &= \{P_t(t)|P_{s,i-1} \leq P_t(t) < P_{s,i}\}\\ \mathbb{P}_{t,M_a} &= \{P_t(t)|P_{s,M_a-1} \leq P_t(t)\}\end{aligned} \tag{5.3}$$

where $P_{s,1} < P_{s,2} < ... < P_{s,i} < ... < P_{s,M_a-1}$ and all $P_{s,i}$ are constants for $i \in [1, 2, ..., M_a - 1]$.

In order to achieve good operation reliability and a long lifespan the temperature should be controlled to keep it in an appropriate temperature range. As emphasized in [VWWL00] a temperature increase can strongly degrade the chip lifespan. The appropriate temperature range for a chip depends on the design of the MCP and is independent of the target power. A common way to control the temperature of a MCP is to ensure that maximum temperature of the dies does not exceed a boundary temperature, see [FKLK12].

Based on this idea the maximum temperature among all die blocks is considered for the velocity adjustment policy. The maximum temperature among all die blocks is given by

$$T_m(t) = \max_{j=\{1,2,...,M_1\}} T_{d,j}(t), \tag{5.4}$$

where M_1 is the number of blocks in the die, see Section 4.3. Further, we define possible regions for maximum temperature $T_m(t)$ given by M_b sets

$$\begin{aligned}\mathbb{T}_{m,1} &= \{T_m(t)|T_m(t) < T_{m,1}\}\\ \mathbb{T}_{m,i} &= \{T_m(t)|T_{m,i-1} \leq T_m(t) < T_{m,i}\}\\ \mathbb{T}_{m,M_b} &= \{T_m(t)|T_{m,M_b-1} \leq T_m(t)\}\end{aligned} \tag{5.5}$$

where $T_{m,1} < T_{m,2} < ... < T_{m,i} < ... < T_{m,M_b-1}$. The aim of the micro-channel cooling system is to reduce the temperature, where basically a further temperature decrease can be achieved by a liquid velocity increase. However, in this approach we also take the energy consumption of the cooling system into account. Therefore we define $\mathbb{T}_{m,1}$ as the 'too low' temperature area, i.e. if $T_m(t) \in \mathbb{T}_{m,1}$, the pump can adjust to a lower fluid velocity which can save energy. The second set $\mathbb{T}_{m,2}$ is then defined as the deserved temperature area. The aim is to keep the temperature of dies in this area under the control policy proposed in Section 5.2.2. Similar to $u_{in,i}$, the difference between each adjacent $T_{m,i}$ can be set as a constant but not necessarily for $3 \leq i \leq M_b$. As the deserved temperature area does not depend on the target power the sets $T_{m,i}$ are fixed independent of the target power. By this fluid velocity adjustment policy, a trade-off between energy consumption and chip lifespan is achieved (mainly by the definition of $\mathbb{T}_{m,1}$ and $\mathbb{T}_{m,2}$).

Remark 5.1. *The domain of each set $\mathbb{P}_{t,i}$ and the fluid velocity $u_{in,i}$ should be designed jointly such that under an equivalent target power of each core the control policy proposed in Section 5.2.2 keeps the maximum temperature in the deserved area $T_m(t) \in \mathbb{T}_{m,2}$. However, it is also possible to define a varying deserved temperature area with respect to the workload. Therefore, the following logic algorithm 5.1 needs to be slightly adapted.*

The liquid velocity adjustment logic algorithm is described in the Algorithm 5.1 . Here we consider the system is digitally controlled with the sampling period t_s, and set the initial time $t_0 = 0$ and k is the kth sampling instant. The basic ideal of the algorithm is to adjust the fluid velocity $u_{\text{in}}(k)$ based on the target power consumption and the maximum temperature of the dies. The variable l_s measures the amount of sampling periods that T_{m} is kept in the set $\mathbb{T}_{\text{m},i}$ with $i \neq 2$. Further, we define the constant $l_c = \left\lfloor \frac{t_c}{t_s} \right\rfloor$, where t_c is the largest time constant of all the subsystems defined in Section 5.2.1.

Remark 5.2. *In the algorithm 5.1 $T_m(k)$ and $T_m(k-1)$ are compared. If $T_m(k-1)$ is higher then $T_m(k)$, that indicates that the temperature is already decreasing under the velocity $u_{in}(k)$, therefore $u_{in}(k)$ should keep the value of the previous step. Otherwise, it will cause oscillation of the velocity input $u_{in}(k)$ and $T_m(k)$ on the two set boundary. If $T_m(k)$ stays for a longer time than τ in $\mathbb{T}_{m,i}$ with $i > 2$, it demonstrates that the fluid velocity is not big enough to set $T_m(k)$ back to $\mathbb{T}_{m,2}$, therefore, $u_{in}(k)$ should be increased.*

5.2 Control design for balancing

Based on logic fluid velocity adjusting law developed in the last section, for the fluid velocity $u_{\text{in},\delta(k)}$, the system (4.48) can be rewritten as

$$\dot{\boldsymbol{T}}(t) = \boldsymbol{A}(\boldsymbol{T}(t), u_{\text{in},\delta(k)})\boldsymbol{T}(t) + \boldsymbol{B}_1\boldsymbol{P}(t) + \boldsymbol{B}_2\boldsymbol{d}(t) + \boldsymbol{B}_3 T_\infty + \boldsymbol{B}_4 T_{\text{in}} \tag{5.6}$$

In this model, the system state matrix $\boldsymbol{A}(\boldsymbol{T}(t), u_{\text{in},\delta(k)})$ is affected by the fluid flow $u_{\text{in}}(t)$ and the temperature $\boldsymbol{T}(t)$. According to the simulation result shown in Figure 4.18, the variation of $\boldsymbol{T}_{\text{f}}(t)$ will lead to a minor variation of the forced convective heat transfer coefficients. Consequently, under the variation range of the liquid temperature $\boldsymbol{T}_{\text{f}}(t)$, the variation of the convective heat transfer coefficient h_{f} in a given block and under a given liquid velocity $u_{\text{in},\delta(k)}$ is very small. Therefore, we define an average value of the convective heat transfer coefficient

$$\bar{h}_{\text{f},\delta(k),n} = \frac{\max(h_{\text{f},\delta(k),n}) + \min(h_{\text{f},\delta(k),n})}{2}, \tag{5.7}$$

where $\max(h_{\text{f},\delta(k),n})$ and $\min(h_{\text{f},\delta(k),n})$ is the largest and smallest heat convection coefficient in a given channel block n and under a given liquid velocity. We know that in different blocks the convective heat transfer coefficient differs, which is considered in the elements of the system matrix, see Chapter 4. With this approximation, the time-varying influence on the system matrix is eliminated such that $\boldsymbol{A}_{\delta(k)}$ is constant for a given switching index $\delta(k)$.

The measurement output is defined as

$$\boldsymbol{\xi}_1(t) = \boldsymbol{C}_{\text{o}}\boldsymbol{T}(t) \tag{5.8}$$

Algorithm 5.1 The switching law

Input: $\boldsymbol{\xi}_1(k)$, $P_t(k)$
Output: $\delta(k)$

Set $\boldsymbol{P}_{\mathrm{t}}(-1) = 0$.
while 1 **do**
  Get $P_{\mathrm{t}}(k)$ and $\boldsymbol{\xi}_1(k)$
  if $P_{\mathrm{t}}(k) \neq P_{\mathrm{t}}(k-1)$ **then**
    $\delta(k) = i, \quad \text{if} \quad P_t \in \mathbb{P}_{\mathrm{t},i}$
    $u_{\mathrm{in}}(k) = u_{\mathrm{in},\delta}$
  end if
  Get $\widehat{\boldsymbol{T}}(k)$, get T_{m} // $\widehat{\boldsymbol{T}}(k)$ *is the estimated state shown in 5.2.1*
  if $T_{\mathrm{m}} \notin \mathbb{T}_{\mathrm{m},2}$ **then**
    $l_{\mathrm{s}} = l_{\mathrm{s}} + 1$
    if $T_m \in \mathbb{T}_{\mathrm{m},1}$ **then**
      if $\Big(\delta(k-1) > 1 \text{ and } l_{\mathrm{s}} = 0\Big)$ or $(l_{\mathrm{s}} > l_{\mathrm{c}})$ **then**
        $\delta(k) = \delta(k-1) - 1$
        $l_{\mathrm{s}} = l_{\mathrm{s}} + 1$
      else
        $\delta(k) = \delta(k-1)$
      end if
    else
      find out i for $T_{\mathrm{m}}(k) \in \mathbb{T}_{\mathrm{m},i}$
      if $\Big((T_{\mathrm{m}}(k-1) \in \mathbb{T}_{\mathrm{t},i}) \text{ or } (T_{\mathrm{m}}(k) < T_{\mathrm{m}}(k-1))\Big)$ and $(l_{\mathrm{s}} < l_{\mathrm{c}})$ **then**
        $\delta(k) = \delta(k-1)$
        $l_{\mathrm{s}} = l_{\mathrm{s}} + 1$
      else
        $\delta(k) = \delta(k-1) + i - 2$ // -2 *is because of optimal set has the index 2*
        $l_{\mathrm{s}} = 0$
      end if
    end if
  else
    $\delta(k) = \delta(k-1)$
    $l_{\mathrm{s}} = 0$
  end if
  update $\boldsymbol{P}_{\mathrm{c}}(t)$ // $\boldsymbol{P}_c(t)$ *is the controller output shown in 5.2.2*
end while

$\boldsymbol{C}_{\mathrm{o}}$ is the measurement matrix and $\boldsymbol{C}_{\mathrm{o}} \in \mathbb{R}^{M_{\mathrm{c}}\times(M_1+M_2)}$, where M_{c} is the number of digital thermal sensors (DTS). Assume each core block contains one DTS, and some other die blocks may also contain DTSs. The DTS planning depends on the layout of the dies [ZAD13].

The regulated output is defined as the temperature difference among cores, i.e.

$$\boldsymbol{\xi}_2(t) = \boldsymbol{C}_{\mathrm{r}}\boldsymbol{T}(t) \tag{5.9}$$

where

$$\xi_{2,i}(t) = T_{\mathrm{d},i}(t) - \sum_{j=1}^{N} T_{\mathrm{d},j}(t) \qquad i = 1, 2, ..., N, \tag{5.10}$$

$$\boldsymbol{C}_{\mathrm{r}} = \begin{bmatrix} \frac{N-1}{N} & -\frac{1}{N} & \cdots & -\frac{1}{N} & 0 & \cdots & 0 \\ -\frac{1}{N} & \frac{N-1}{N} & \cdots & -\frac{1}{N} & 0 & \cdots & 0 \\ \vdots & \vdots & \vdots & \vdots & \vdots & \vdots & \vdots \\ -\frac{1}{N} & \cdots & -\frac{1}{N} & \frac{N-1}{N} & 0 & \cdots & 0 \end{bmatrix}.$$

and $\boldsymbol{C}_r \in \mathbb{R}^{N\times(M_1+M_2)}$ is the regulated output matrix.

Discretize the system (5.6) as

$$\boldsymbol{T}(k+1) = \boldsymbol{A}_{\mathrm{d},\delta(k)}\boldsymbol{T}(k) + \boldsymbol{B}_{\mathrm{d1},\delta(k)}\boldsymbol{P}(k) + \boldsymbol{B}_{\mathrm{d3},\delta(k)}T_{\infty} + \boldsymbol{B}_{\mathrm{d4},\delta(k)}T_{\mathrm{in}} + \boldsymbol{d}(k), \tag{5.11}$$

where

$$\begin{aligned} \boldsymbol{A}_{\mathrm{d},\delta(k)} &= e^{\boldsymbol{A}_{\delta(k)}t_{\mathrm{s}}}, \\ \boldsymbol{B}_{\mathrm{d1},\delta(k)} &= \int_0^{t_{\mathrm{s}}} e^{\boldsymbol{A}_{\delta(k)}t} dt \boldsymbol{B}_1, \\ \boldsymbol{B}_{\mathrm{d3},\delta(k)} &= \int_0^{t_{\mathrm{s}}} e^{\boldsymbol{A}_{\delta(k)}t} dt \boldsymbol{B}_3, \\ \boldsymbol{B}_{\mathrm{d4},\delta(k)} &= \int_0^{t_{\mathrm{s}}} e^{\boldsymbol{A}_{\delta(k)}t} dt \boldsymbol{B}_4, \\ \boldsymbol{d}(k) &= \int_{k\cdot t_{\mathrm{s}}}^{(k+1)\cdot t_{\mathrm{s}}} e^{\boldsymbol{A}_{\delta(k)}t}\boldsymbol{B}_2\boldsymbol{d}(t)dt. \end{aligned}$$

The system measurement output and regulated output are

$$\boldsymbol{\xi}_1(k) = \boldsymbol{C}_{\mathrm{o}}\boldsymbol{T}(k) \tag{5.12}$$

$$\boldsymbol{\xi}_2(k) = \boldsymbol{C}_r\boldsymbol{T}(k). \tag{5.13}$$

5.2.1 Robust H_∞ observer design

In Section 5.2.2 an MPC controller is designed for the balancing for which all states $\boldsymbol{T}(t)$ are required. However, only $\boldsymbol{\xi}_1(t)$ is measurable. The other states, for instance the temperature of the liquid $\boldsymbol{T}_{\mathrm{f}}(t)$ and of some of the die blocks $T_{\mathrm{d},i}(t)$ are not measurable depending on the DTS planing. Therefore an observer is required to estimate the states. In the observer design it needs to be considered that the states are affected by a bounded nonnegative disturbance vector, see equation (5.11).

The disturbance is generally not constant and its dynamics can usually not be determined as the disturbance is affected by many parts of the MCP as for instance the I/O controller, the L2 cache and the interconnect bus which work independently. Further, the boundedness of the disturbance is small considering the influence on the estimation. For some die blocks the influence of the disturbance is negligible due to their position in the MCP with respect to the disturbing parts.

Therefore, a robust H_∞ current state observer is introduced to limit the influence of the unknown disturbance on the estimation. The observer consists of two parts.

1. The prediction of the state vector is

$$\begin{aligned}\overline{\boldsymbol{T}}(k) =& \boldsymbol{A}_{\mathrm{d},\delta(k-1)}\widehat{\boldsymbol{T}}(k-1) + \boldsymbol{B}_{\mathrm{d1},\delta(k-1)}(k-1)\boldsymbol{P}(k-1) \\ &+ \boldsymbol{B}_{\mathrm{d3},\delta(k-1)}(k-1)T_\infty + \boldsymbol{B}_{\mathrm{d4},\delta(k-1)}T_{\mathrm{in}}.\end{aligned} \tag{5.14}$$

2. The correction of the predicted state vector based on the actual measurements is

$$\widehat{\boldsymbol{T}}(k) = \overline{\boldsymbol{T}}(k) + \boldsymbol{L}_{\delta(k-1)}\big(\boldsymbol{\xi}_1(k) - \overline{\boldsymbol{\xi}}_1(k)\big) \tag{5.15}$$

where $\overline{\boldsymbol{T}}(k) \in \mathbb{R}^{M_1+M_2}$ is predicted state vector

$$\overline{\boldsymbol{\xi}}_1(k) = \boldsymbol{C}_{\mathrm{o}}\overline{\boldsymbol{T}}(k). \tag{5.16}$$

$\widehat{\boldsymbol{T}}(k) \in \mathbb{R}^{M_1+M_2}$ is the corrected state vector

$$\widehat{\boldsymbol{\xi}}_1(k) = \boldsymbol{C}_{\mathrm{o}}\widehat{\boldsymbol{T}}(k), \tag{5.17}$$

and $\boldsymbol{L}_{\delta(k-1)} \in \mathbb{R}^{(M_1+M_2)\times N}$ is the observer gain matrix.

Define the estimation error as

$$\boldsymbol{e}(k) = \boldsymbol{T}(k) - \widehat{\boldsymbol{T}}(k). \tag{5.18}$$

Therefore, the error dynamics can be gotten by subtracting (5.15) from (5.11)

$$\begin{aligned}\boldsymbol{e}(k) =& \boldsymbol{T}(k) - \widehat{\boldsymbol{T}}(k) \\ =& \boldsymbol{A}_{\mathrm{d},\delta(k-1)}\boldsymbol{e}(k-1) + \boldsymbol{d}(k-1) - \boldsymbol{L}_{\delta(k-1)}\boldsymbol{C}_{\mathrm{o}}\Big(\boldsymbol{T}(k) - \overline{\boldsymbol{T}}(k)\Big).\end{aligned} \tag{5.19}$$

Consider that

$$T(k) - \overline{T}(k) = A_{d,\delta(k-1)}\Big(T(k-1) - \widehat{T}(k-1)\Big) + d(k-1), \tag{5.20}$$

one has

$$\begin{aligned} e(k) =& A_{d,\delta(k-1)}(T(k-1) - \widehat{T}(k-1)) + d(k-1) \\ &- L_{\delta(k-1)}C_o(A_{d,\delta(k-1)}(T(k-1) - \widehat{T}(k-1)) + d(k-1)) \\ =& A_{d,\delta(k-1)}e(k-1) - L_{\delta(k-1)}C_oA_{d,\delta(k-1)}e(k-1) + (I - L_{\delta(k-1)}C_o)d(k-1) \\ =& (A_{d,\delta(k-1)} - L_{\delta(k-1)}C_oA_{d,\delta(k-1)})e(k-1) + (I - L_{\delta(k-1)}C_o)d(k-1) \end{aligned} \tag{5.21}$$

To minimize the influence of the disturbance vector $d(k)$ on the estimation error $e(k)$, an H_∞ observer is designed. The $\mathscr{L}_2$-gain between the disturbance and the estimation error is defined as

$$\gamma = \sup_{\|d(k)\|_2 \neq 0} \frac{\|e(k)\|_2}{\|d(k)\|_2} \tag{5.22}$$

where γ is a positive scalar. By the definition in [ZDG06], (5.22) can be rewritten as

$$\sum_{j=0}^{k} e^T(j)e(j) - \gamma^2 \sum_{j=0}^{k} d^T(j)d(j) \leq 0. \tag{5.23}$$

In the following, the $\mathscr{L}_2$-gain which is defined by the H_∞ norm of the transfer function of a linear time-invariant system (LTI) is be minimized. The H_∞ robust observer (5.14) can be achieved by the following theorem.

Theorem 5.1. *For the system (5.11) with the observer (5.14) and (5.15), the robust H_∞ observer gain L_δ can be achieved by solving the LMI optimiztion problem*

$$\min \ \gamma^2 \tag{5.24a}$$

$$\text{subject to } \mathscr{P} = \mathscr{P}^T > 0, \quad \gamma > 0 \tag{5.24b}$$

$$\begin{bmatrix} \mathscr{P} - I & * & * \\ \mathscr{P}A_{d,\delta} - W_\delta C_o A_{d,\delta} & \mathscr{P} & * \\ 0 & (\mathscr{P} - W_\delta C_o)^T & \gamma^2 I \end{bmatrix} > 0. \tag{5.24c}$$

for all $\delta \in \{1, ..., M_a\}$, where $*$ denotes the entries induced by symmetry. Thus, the observer is constructed by

$$L_\delta = \mathscr{P}^{-1} W_\delta. \tag{5.25}$$

Proof. Consider the Lyapunov function candidate for the error dynamics system (5.19)

$$V(e(k)) = e^T(k)\mathscr{P}e(k) \tag{5.26}$$

where $\mathscr{P} \in \mathbb{R}^{(M1+M2)\times(M1+M2)}$ is symmetric and positive definite. Further, we suppose the follow inequality is satisfied

$$V(e(k+1)) - V(e(k)) + e^T(k)e(k) - \gamma^2 d^T(k)d(k) < 0 \tag{5.27}$$

for all the $\boldsymbol{e}(k)$ and $\boldsymbol{d}(k)$. According to (5.21)

$$
\begin{aligned}
V(\boldsymbol{e}(k+1)) =& \boldsymbol{e}^T(k)\big(\boldsymbol{A}_{\mathrm{d},\delta(k)} - \boldsymbol{L}_{\delta(k)}\boldsymbol{C}_{\mathrm{o}}\boldsymbol{A}_{\mathrm{d},\delta(k)}\big)^T \mathscr{P}\big(\boldsymbol{A}_{\mathrm{d},\delta(k)} - \boldsymbol{L}_{\delta(k)}\boldsymbol{C}_{\mathrm{o}}\boldsymbol{A}_{\mathrm{d},\delta(k)}\big)\boldsymbol{e}(k) \\
&+ \boldsymbol{e}^T(k)\big(\boldsymbol{A}_{\mathrm{d},\delta(k)} - \boldsymbol{L}_{\delta(k)}\boldsymbol{C}_{\mathrm{o}}\boldsymbol{A}_{\mathrm{d},\delta(k)}\big)^T \mathscr{P}\big(\boldsymbol{I} - \boldsymbol{L}_{\delta(k)}\boldsymbol{C}_{\mathrm{o}}\big)\boldsymbol{d}(k) \\
&\boldsymbol{d}^T(k)\big(\boldsymbol{I} - \boldsymbol{L}_{\delta(k)}\boldsymbol{C}_{\mathrm{o}}\big)^T \mathscr{P}\big(\boldsymbol{A}_{\mathrm{d},\delta(k)} - \boldsymbol{L}_{\delta(k)}\boldsymbol{C}_{\mathrm{o}}\boldsymbol{A}_{\mathrm{d},\delta(k)}\big)\boldsymbol{e}(k) \\
&\boldsymbol{d}^T(k)\big(\boldsymbol{I} - \boldsymbol{L}_{\delta(k)}\boldsymbol{C}_{\mathrm{o}}\big)^T \mathscr{P}\big(\boldsymbol{I} - \boldsymbol{L}_{\delta(k)}\boldsymbol{C}_{\mathrm{o}}\big)\boldsymbol{d}(k) \\
=& \begin{bmatrix} \boldsymbol{e}(k) \\ \boldsymbol{d}(k) \end{bmatrix}^T \begin{bmatrix} \tilde{\boldsymbol{A}}^T_{\delta(k)}\mathscr{P}\tilde{\boldsymbol{A}}_{\delta(k)} & * \\ \boldsymbol{H}^T_{\delta(k)}\mathscr{P}\tilde{\boldsymbol{A}}_{\delta(k)} & \boldsymbol{H}^T_{\delta(k)}\mathscr{P}\boldsymbol{H}_{\delta(k)} \end{bmatrix} \begin{bmatrix} \boldsymbol{e}(k) \\ \boldsymbol{d}(k) \end{bmatrix},
\end{aligned}
\tag{5.28}
$$

where $\tilde{\boldsymbol{A}}_{\delta(k)} = \boldsymbol{A}_{\mathrm{d},\delta(k)} - \boldsymbol{L}_{\delta(k)}\boldsymbol{C}_{\mathrm{o}}\boldsymbol{A}_{\mathrm{d},\delta(k)}$ and $\boldsymbol{H}_{\delta(k)} = \boldsymbol{I} - \boldsymbol{L}_{\delta(k)}\boldsymbol{C}_{\mathrm{o}}$. Therefore, (5.27) can equivalently be written as

$$
\begin{bmatrix} \boldsymbol{e}(k) \\ \boldsymbol{d}(k) \end{bmatrix}^T \begin{bmatrix} \tilde{\boldsymbol{A}}^T_{\delta(k)}\mathscr{P}\tilde{\boldsymbol{A}}_{\delta(k)} - \mathscr{P} + \boldsymbol{I} & * \\ \boldsymbol{H}^T_{\delta(k)}\mathscr{P}\tilde{\boldsymbol{A}}_{\delta(k)} & \boldsymbol{H}^T_{\delta(k)}\mathscr{P}\boldsymbol{H}_{\delta(k)} - \gamma^2\boldsymbol{I} \end{bmatrix} \begin{bmatrix} \boldsymbol{e}(k) \\ \boldsymbol{d}(k) \end{bmatrix} < 0, \tag{5.29}
$$

which is satisfied if

$$
\begin{bmatrix} \mathscr{P} - \boldsymbol{I} - \tilde{\boldsymbol{A}}^T_{\delta}\mathscr{P}\tilde{\boldsymbol{A}}_{\delta} & * \\ -\boldsymbol{H}^T_{\delta}\mathscr{P}\tilde{\boldsymbol{A}}_{\delta} & \gamma^2\boldsymbol{I} - \boldsymbol{H}^T_{\delta}\mathscr{P}\boldsymbol{H}_{\delta} \end{bmatrix} > 0 \tag{5.30}
$$

for all $\delta \in \{1, ..., M_{\mathrm{a}}\}$. For simplicity, the time dependency of the switching index is omitted in the remaining part of the proof. By applying the Schur complement A.8, (5.30) equals to the following two conditions

$$
\mathscr{P} - \boldsymbol{I} - \tilde{\boldsymbol{A}}^T_{\delta}\mathscr{P}\tilde{\boldsymbol{A}}_{\delta} - \tilde{\boldsymbol{A}}^T_{\delta}\mathscr{P}\boldsymbol{H}_{\delta}(\gamma^2\boldsymbol{I} - \boldsymbol{H}^T_{\delta}P\boldsymbol{H}_{\delta})^{-1}\boldsymbol{H}^T_{\delta}\mathscr{P}\tilde{\boldsymbol{A}}_{\delta} > 0, \tag{5.31a}
$$

$$
\gamma^2\boldsymbol{I} - \boldsymbol{H}^T_{\delta}\mathscr{P}\boldsymbol{H}_{\delta} > 0. \tag{5.31b}
$$

According to the matrix inversion Lemma A.9, (5.31a) can be rewritten as

$$
\mathscr{P} - \boldsymbol{I} - \tilde{\boldsymbol{A}}^T_{\delta}\mathscr{P}\tilde{\boldsymbol{A}}_{\delta} - \tilde{\boldsymbol{A}}^T_{\delta}\left(\mathscr{P}^{-1} - \frac{1}{\gamma^2}\boldsymbol{H}_{\delta}\boldsymbol{H}^T_{\delta}\right)^{-1}\tilde{\boldsymbol{A}}_{\delta} + \tilde{\boldsymbol{A}}^T_{\delta}\mathscr{P}\tilde{\boldsymbol{A}}_{\delta} > 0. \tag{5.32}
$$

As $\mathscr{P}$ is positive definite the Schur complement is applied again to condition (5.31b) resulting in

$$
\begin{bmatrix} \gamma^2\boldsymbol{I} & \boldsymbol{H}^T_{\delta} \\ \boldsymbol{H}_{\delta} & \mathscr{P}^{-1} \end{bmatrix} > 0. \tag{5.33}
$$

This shows that (5.31) is equivalent to

$$
\mathscr{P}^{-1} - \frac{1}{\gamma^2}\boldsymbol{H}_{\delta}\boldsymbol{H}^T_{\delta} > 0 \tag{5.34a}
$$

$$
\mathscr{P} - \boldsymbol{I} - \tilde{\boldsymbol{A}}^T_{\delta}\left(\mathscr{P}^{-1} - \frac{1}{\gamma^2}\boldsymbol{H}_{\delta}\boldsymbol{H}^T_{\delta}\right)^{-1}\tilde{\boldsymbol{A}}_{\delta} > 0 \tag{5.34b}
$$

Applying the Schur complement, one has

$$\begin{bmatrix} \mathscr{P} - \boldsymbol{I} & * \\ \tilde{\boldsymbol{A}}_\delta & \mathscr{P}^{-1} - \frac{1}{\gamma^2}\boldsymbol{H}_\delta \boldsymbol{H}_\delta^T \end{bmatrix} > 0, \tag{5.35}$$

and (5.35) can be written as

$$\begin{bmatrix} \mathscr{P} - \boldsymbol{I} & * \\ \tilde{\boldsymbol{A}}_\delta & \mathscr{P}^{-1} \end{bmatrix} - \begin{bmatrix} 0 \\ \boldsymbol{H}_\delta \end{bmatrix} \frac{1}{\gamma^2}\boldsymbol{I} \begin{bmatrix} 0 & \boldsymbol{H}_\delta^T \end{bmatrix} > 0. \tag{5.36}$$

Applying the Schur complement, (5.36) is transformed to

$$\begin{bmatrix} \mathscr{P} - \boldsymbol{I} & * & * \\ \tilde{\boldsymbol{A}}_\delta & \mathscr{P}^{-1} & * \\ 0 & \boldsymbol{H}_\delta^T & \gamma^2 \boldsymbol{I} \end{bmatrix} > 0. \tag{5.37}$$

Pre- and post-multiplying (5.37) by block-diagonal matrix $\mathrm{diag}(\boldsymbol{I}, \mathscr{P}, \boldsymbol{I})$, we have

$$\begin{bmatrix} \mathscr{P} - \boldsymbol{I} & * & * \\ \mathscr{P}\tilde{\boldsymbol{A}}_\delta & \mathscr{P} & * \\ 0 & \boldsymbol{H}_\delta^T \mathscr{P} & \gamma^2 \boldsymbol{I} \end{bmatrix} > 0. \tag{5.38}$$

Defining $\mathscr{P}\boldsymbol{L}_\delta = \boldsymbol{W}_\delta$, (5.38) can be rewritten as

$$\begin{bmatrix} \mathscr{P} - \boldsymbol{I} & * & * \\ \mathscr{P}\boldsymbol{A}_{\mathrm{d},\delta} - \boldsymbol{W}_\delta \boldsymbol{C}_{\mathrm{o}} \boldsymbol{A}_{\mathrm{d},\delta} & \mathscr{P} & * \\ 0 & (\mathscr{P} - \boldsymbol{W}_\delta \boldsymbol{C}_{\mathrm{o}})^T & \gamma^2 \boldsymbol{I} \end{bmatrix} > 0. \tag{5.39}$$

This completes the proof. □

As a common Lyapunov function is utilized for the observer design and the disturbance is bounded (4.49), $\mathscr{L}_2$-gain stability is guaranteed under an arbitrary switching sequence given by Algorithm 5.1.

5.2.2 MPC controller design

As mentioned in Chapter 3, in order to reach the processor task target, the power is divided into two parts, i.e.

$$\boldsymbol{P}(k) = \boldsymbol{P}_{\mathrm{c}}(k) + \boldsymbol{P}_{\mathrm{t}}(k) \tag{5.40}$$

where $\boldsymbol{P}_{\mathrm{t}}(k)$ is the core power based on the target tasks, and $\boldsymbol{P}_{\mathrm{c}}(k)$ is the controller output.

Suppose l_1 and l_2 are the control horizon and prediction horizon and define $l_1 \leq l_2$ [CB07, Ch. 2]. Then the prediction of system states under the switching index δ is

$$\begin{aligned}\boldsymbol{T}(k+l|k) =& \boldsymbol{A}_{\mathrm{d},\delta}\boldsymbol{T}(k+l-1|k) + \boldsymbol{B}_{\mathrm{d1},\delta}\boldsymbol{P}(k+l-1) + \boldsymbol{B}_{\mathrm{d3},\delta}T_{\infty} + \boldsymbol{B}_{\mathrm{d4},\delta}T_{\mathrm{in}} \\ =& \boldsymbol{A}^2_{\mathrm{d},\delta}\boldsymbol{T}(k+l-2|k) + \boldsymbol{A}_{\mathrm{d},\delta}\boldsymbol{B}_{\mathrm{d1},\delta}\boldsymbol{P}(k+l-2) + \boldsymbol{A}_{\mathrm{d},\delta}\boldsymbol{B}_{\mathrm{d3},\delta}T_{\infty} \\ &+ \boldsymbol{A}_{\mathrm{d},\delta}\boldsymbol{B}_{\mathrm{d4},\delta}T_{\mathrm{in}} + \boldsymbol{B}_{\mathrm{d1},\delta}\boldsymbol{P}(k+l-1) + \boldsymbol{B}_{\mathrm{d3},\delta}T_{\infty} + \boldsymbol{B}_{\mathrm{d4},\delta}T_{\mathrm{in}} \\ =& \ldots \\ =& \boldsymbol{A}^l_{\mathrm{d},\delta}\boldsymbol{T}(k) + \sum_{j=1}^{l}\boldsymbol{A}^{j-1}_{\mathrm{d},\delta}\boldsymbol{B}_{\mathrm{d1},\delta}\boldsymbol{P}(k+l-j) + \sum_{j=1}^{l}\boldsymbol{A}^{j-1}_{\mathrm{d}i}\boldsymbol{B}_{\mathrm{d3},\delta}T_{\infty} \\ &+ \sum_{j=1}^{l}\boldsymbol{A}^{j-1}_{\mathrm{d},\delta}\boldsymbol{B}_{\mathrm{d4},\delta}T_{\mathrm{in}}\end{aligned} \tag{5.41}$$

and the regulated output is

$$\begin{aligned}\boldsymbol{\xi}_2(k+l|k) =& \boldsymbol{C}_{\mathrm{r}}\boldsymbol{T}(k+l|k) \\ =& \boldsymbol{C}_{\mathrm{r}}\boldsymbol{A}^l_{\mathrm{d}i}\boldsymbol{T}(k) + \boldsymbol{C}_{\mathrm{r}}\sum_{j=1}^{l}\boldsymbol{A}^{j-1}_{\mathrm{d}i}\boldsymbol{B}_{1d}\boldsymbol{P}(k+l-j) \\ &\times \boldsymbol{C}_{\mathrm{r}}\sum_{j=1}^{l}\boldsymbol{A}^{j-1}_{\mathrm{d}i}\boldsymbol{B}_{3d}T_{\infty} + \boldsymbol{C}_{\mathrm{r}}\sum_{j=1}^{l}\boldsymbol{A}^{j-1}_{\mathrm{d}i}\boldsymbol{B}_{4d}T_{\mathrm{in}}\end{aligned} \tag{5.42}$$

Therefore, the whole prediction regulated output from $k+1$ to $k+l_2$ is

$$\widehat{\boldsymbol{\xi}}_2 = \widehat{\boldsymbol{A}}_{\mathrm{d},\delta}\boldsymbol{T}(k) + \widehat{\boldsymbol{B}}_{\mathrm{d1},\delta}\widehat{\boldsymbol{P}} + \widehat{\boldsymbol{B}}_{\mathrm{d3},\delta}T_{\infty} + \widehat{\boldsymbol{B}}_{\mathrm{d4},\delta}T_{\mathrm{in}} \tag{5.43}$$

where

$$\widehat{\boldsymbol{\xi}}_2 = \begin{bmatrix}\boldsymbol{\xi}_2(k+1|k) \\ \boldsymbol{\xi}_2(k+2|k) \\ \ldots \\ \boldsymbol{\xi}_2(k+l_1|k) \\ \ldots \\ \boldsymbol{\xi}_2(k+l_2|k)\end{bmatrix} \qquad \widehat{\boldsymbol{P}} = \begin{bmatrix}\boldsymbol{P}(k) \\ \boldsymbol{P}(k+1) \\ \ldots \\ \boldsymbol{P}(k+l_1-1)\end{bmatrix}, \tag{5.44}$$

and

$$\widehat{\boldsymbol{A}}_{\mathrm{d},\delta} = \begin{bmatrix}\boldsymbol{C}_{\mathrm{r}}\boldsymbol{A}_{\mathrm{d},\delta} & \boldsymbol{C}_{\mathrm{r}}\boldsymbol{A}^2_{\mathrm{d}i} & \ldots & \boldsymbol{C}_{\mathrm{r}}\boldsymbol{A}^{l_1}_{\mathrm{d}i} & \ldots & \boldsymbol{C}_{\mathrm{r}}\boldsymbol{A}^{l_2}_{\mathrm{d}i}\end{bmatrix}^T, \tag{5.45}$$

$$\widehat{\boldsymbol{B}}_{\mathrm{d1},\delta} = \begin{bmatrix}\boldsymbol{C}_{\mathrm{r}}\boldsymbol{B}_{\mathrm{d1},\delta} & 0 & 0 & \ldots & 0 \\ \boldsymbol{C}_{\mathrm{r}}\boldsymbol{A}_{\mathrm{d},\delta}\boldsymbol{B}_{\mathrm{d1},\delta} & \boldsymbol{C}_{\mathrm{r}}\boldsymbol{B}_{\mathrm{d1},\delta} & 0 & \ldots & 0 \\ \ldots & & & & \\ \boldsymbol{C}_{\mathrm{r}}\boldsymbol{A}^{l_1-1}_{\mathrm{d},\delta}\boldsymbol{B}_{\mathrm{d1},\delta} & \boldsymbol{C}_{\mathrm{r}}\boldsymbol{A}^{l_1-2}_{\mathrm{d},\delta}\boldsymbol{B}_{\mathrm{d1},\delta} & \boldsymbol{C}_{\mathrm{r}}\boldsymbol{A}^{l_1-3}_{\mathrm{d},\delta}\boldsymbol{B}_{\mathrm{d1},\delta} & \ldots & \boldsymbol{C}_{\mathrm{r}}\boldsymbol{B}_{\mathrm{d1},\delta} \\ \ldots & & & & \\ \boldsymbol{C}_{\mathrm{r}}\boldsymbol{A}^{l_2-1}_{\mathrm{d},\delta}\boldsymbol{B}_{\mathrm{d1},\delta} & \boldsymbol{C}_{\mathrm{r}}\boldsymbol{A}^{l_2-2}_{\mathrm{d},\delta}\boldsymbol{B}_{\mathrm{d1},\delta} & \boldsymbol{C}_{\mathrm{r}}\boldsymbol{A}^{l_2-3}_{\mathrm{d},\delta}\boldsymbol{B}_{\mathrm{d1},\delta} & \ldots & \boldsymbol{C}_{\mathrm{r}}\boldsymbol{A}^{l_2-l_1}_{\mathrm{d},\delta}\boldsymbol{B}_{\mathrm{d1},\delta}\end{bmatrix},$$

$$\widehat{\boldsymbol{B}}_{\mathrm{d3},\delta} = \begin{bmatrix} \boldsymbol{C}_{\mathrm{r}}\boldsymbol{B}_{\mathrm{d3},\delta} & 0 & 0 & \dots & 0 \\ \boldsymbol{C}_{\mathrm{r}}\boldsymbol{A}_{\mathrm{d},\delta}\boldsymbol{B}_{\mathrm{d3},\delta} & \boldsymbol{C}_{\mathrm{r}}\boldsymbol{B}_{\mathrm{d3},\delta} & 0 & \dots & 0 \\ \dots & & & & \\ \boldsymbol{C}_{\mathrm{r}}\boldsymbol{A}_{\mathrm{d},\delta}^{l_1-1}\boldsymbol{B}_{\mathrm{d3},\delta} & \boldsymbol{C}_{\mathrm{r}}\boldsymbol{A}_{\mathrm{d},\delta}^{l_1-2}\boldsymbol{B}_{\mathrm{d3},\delta} & \boldsymbol{C}_{\mathrm{r}}\boldsymbol{A}_{\mathrm{d},\delta}^{l_1-3}\boldsymbol{B}_{\mathrm{d3},\delta} & \dots & \boldsymbol{C}_{\mathrm{r}}\boldsymbol{B}_{\mathrm{d3},\delta} \\ \dots & & & & \\ \boldsymbol{C}_{\mathrm{r}}\boldsymbol{A}_{\mathrm{d},\delta}^{l_2-1}\boldsymbol{B}_{\mathrm{d3},\delta} & \boldsymbol{C}_{\mathrm{r}}\boldsymbol{A}_{\mathrm{d},\delta}^{l_2-2}\boldsymbol{B}_{\mathrm{d3},\delta} & \boldsymbol{C}_{\mathrm{r}}\boldsymbol{A}_{\mathrm{d},\delta}^{l_2-3}\boldsymbol{B}_{\mathrm{d3},\delta} & \dots & \boldsymbol{C}_{\mathrm{r}}\boldsymbol{A}_{\mathrm{d},\delta}^{l_2-l_1}\boldsymbol{B}_{\mathrm{d3},\delta} \end{bmatrix} \begin{bmatrix} 1 \\ 1 \\ \dots \\ 1 \\ \dots \\ 1 \end{bmatrix},$$

$$\widehat{\boldsymbol{B}}_{\mathrm{d4},\delta} = \begin{bmatrix} \boldsymbol{C}_{\mathrm{r}}\boldsymbol{B}_{\mathrm{d4},\delta} & 0 & 0 & \dots & 0 \\ \boldsymbol{C}_{\mathrm{r}}\boldsymbol{A}_{\mathrm{d},\delta}\boldsymbol{B}_{\mathrm{d4},\delta} & \boldsymbol{C}_{\mathrm{r}}\boldsymbol{B}_{\mathrm{d4},\delta} & 0 & \dots & 0 \\ \dots & & & & \\ \boldsymbol{C}_{\mathrm{r}}\boldsymbol{A}_{\mathrm{d},\delta}^{l_1-1}\boldsymbol{B}_{\mathrm{d4},\delta} & \boldsymbol{C}_{\mathrm{r}}\boldsymbol{A}_{\mathrm{d},\delta}^{l_1-2}\boldsymbol{B}_{\mathrm{d4},\delta} & \boldsymbol{C}_{\mathrm{r}}\boldsymbol{A}_{\mathrm{d},\delta}^{l_1-3}\boldsymbol{B}_{\mathrm{d4},\delta} & \dots & \boldsymbol{C}_{\mathrm{r}}\boldsymbol{B}_{\mathrm{d4},\delta} \\ \dots & & & & \\ \boldsymbol{C}_{\mathrm{r}}\boldsymbol{A}_{\mathrm{d},\delta}^{l_2-1}\boldsymbol{B}_{\mathrm{d4},\delta} & \boldsymbol{C}_{\mathrm{r}}\boldsymbol{A}_{\mathrm{d},\delta}^{l_2-2}\boldsymbol{B}_{\mathrm{d4},\delta} & \boldsymbol{C}_{\mathrm{r}}\boldsymbol{A}_{\mathrm{d},\delta}^{l_2-3}\boldsymbol{B}_{\mathrm{d4},\delta} & \dots & \boldsymbol{C}_{\mathrm{r}}\boldsymbol{A}_{\mathrm{d},\delta}^{l_2-l_1}\boldsymbol{B}_{\mathrm{d4},\delta} \end{bmatrix} \begin{bmatrix} 1 \\ 1 \\ \dots \\ 1 \\ \dots \\ 1 \end{bmatrix}.$$

The control objective is to minimize the cost function defined by

$$J = \widehat{\boldsymbol{\xi}}_2^T \mathscr{Q} \widehat{\boldsymbol{\xi}}_2 + \widehat{\boldsymbol{P}}_{\mathrm{c}}^T \mathscr{R} \widehat{\boldsymbol{P}}_{\mathrm{c}}, \tag{5.46}$$

where $\mathscr{Q} \in \mathbb{R}^{(M_{\mathrm{c}} \cdot l_2)\times(M_{\mathrm{c}} \cdot l_2)}$ and $\mathscr{R} \in \mathbb{R}^{(N \cdot l_1)\times(N \cdot l_1)}$ are symmetric and positive definite. The first term of the right hand side is linked to the state objective, which is to balance the temperature among the cores, while the second term is related to the management of the power consumption of the cores. To minimize the cost function J, one has

$$\frac{\partial J}{\partial \widehat{\boldsymbol{P}}_{\mathrm{c}}} = 0. \tag{5.47}$$

By substituting (5.43) into (5.46), the optimal solution for the control signal $\widehat{\boldsymbol{P}}_{\mathrm{c}}$ is

$$\widehat{\boldsymbol{P}}_{\mathrm{c}} = -(\widehat{\boldsymbol{B}}_{\mathrm{d1},\delta}^T \mathscr{Q} \widehat{\boldsymbol{B}}_{\mathrm{d1},\delta} + \mathscr{R})^{-1} \widehat{\boldsymbol{B}}_{\mathrm{d1},\delta}^T (\widehat{\boldsymbol{A}}_{\mathrm{d},\delta} \widehat{\boldsymbol{T}}(k) + \widehat{\boldsymbol{B}}_{\mathrm{d3},\delta} T_\infty + \widehat{\boldsymbol{B}}_{\mathrm{d4},\delta} T_{\mathrm{in}}), \tag{5.48}$$

and the control input at the time instant k is

$$\boldsymbol{P}_{\mathrm{c}}(k) = \underbrace{[\boldsymbol{I}, \boldsymbol{0}, \boldsymbol{0}, ..., \boldsymbol{0}]}_{\boldsymbol{I}_{\mathrm{c}}} \widehat{\boldsymbol{P}}_{\mathrm{c}}, \tag{5.49}$$

where $\boldsymbol{I}$ is an $N \times N$ identity matrix. The controller contains three parts, however the parts with T_∞ and T_{in} are constants. Therefore, these two parts will be updated only if the fluid velocity is changed. The controller can equivalently be written as

$$\boldsymbol{P}_{\mathrm{c}}(k) = -\boldsymbol{F}_{\delta(k)} \boldsymbol{T}(k) - \boldsymbol{F}_{\infty,\delta(k)} T_\infty - \boldsymbol{F}_{\mathrm{in},\delta(k)} T_{\mathrm{in}} \tag{5.50}$$

where

$$\boldsymbol{F}_{\delta} = \boldsymbol{I}_{\mathrm{c}} (\widehat{\boldsymbol{B}}_{\mathrm{d1},\delta}^T \mathscr{Q} \widehat{\boldsymbol{B}}_{\mathrm{d1},\delta} + \mathscr{R})^{-1} \widehat{\boldsymbol{B}}_{\mathrm{d1},\delta}^T \widehat{\boldsymbol{A}}_{\mathrm{d},\delta} \tag{5.51a}$$

$$\boldsymbol{F}_{\infty,\delta} = \boldsymbol{I}_{\mathrm{c}} (\widehat{\boldsymbol{B}}_{\mathrm{d1},\delta}^T \mathscr{Q} \widehat{\boldsymbol{B}}_{\mathrm{d1},\delta} + \mathscr{R})^{-1} \widehat{\boldsymbol{B}}_{\mathrm{d1},\delta}^T \widehat{\boldsymbol{B}}_{\mathrm{d3},\delta} \tag{5.51b}$$

$$\boldsymbol{F}_{\mathrm{in},\delta} = \boldsymbol{I}_{\mathrm{c}} (\widehat{\boldsymbol{B}}_{\mathrm{d1},\delta}^T \mathscr{Q} \widehat{\boldsymbol{B}}_{\mathrm{d1},\delta} + \mathscr{R})^{-1} \widehat{\boldsymbol{B}}_{\mathrm{d1},\delta}^T \widehat{\boldsymbol{B}}_{\mathrm{d4},\delta}. \tag{5.51c}$$

As stability is not given inherently for an MPC controller stability needs to be verified. However, the control gains $\boldsymbol{F}_{\delta}$, $\boldsymbol{F}_{\infty,\delta}$ and $\boldsymbol{F}_{\text{in},\delta}$ are constant for a given switching index δ as input and state constraints have not been considered. Further, it is necessary to verify the stability under switching, which is not covered by the control design.

5.2.3 Stability analysis

Based on the given approach a constant controller is designed for each subsystem. Further an H_{∞} observer is designed separately from the control design. As we deal with a switched system, also the validity of the separation principle needs to be analyzed. Therefore, we give the whole switched system model consisting of the closed loop system dynamics and the observation error dynamics, for investigating the stability. As the system is influenced by the unknown input $\boldsymbol{d}(k)$, the target power consumption $\boldsymbol{P}_{\text{t}}(k)$, the fluid inlet temperature T_{in} and the environment temperature T_{∞} we consider the $\mathscr{L}_2$-gain stability. Define a new disturbance input as

$$\begin{aligned}\boldsymbol{d}_{\text{s}}(k) =& \boldsymbol{d}(k) + \boldsymbol{B}_{\text{d1},\delta(k)}\boldsymbol{P}_t(k) + \boldsymbol{B}_{\text{d3},\delta(k)}T_{\infty} + \boldsymbol{B}_{\text{d4},\delta(k)}T_{\text{in}} \\ &+ \boldsymbol{I}_{\text{c}}(\widehat{\boldsymbol{B}}^T_{\text{d1},\delta(k)}\mathscr{Q}\widehat{\boldsymbol{B}}_{\text{d1},\delta(k)} + \mathscr{R})^{-1}\widehat{\boldsymbol{B}}^T_{\text{d1},\delta(k)}(\widehat{\boldsymbol{B}}_{\text{d3},\delta(k)}T_{\infty} + \widehat{\boldsymbol{B}}_{\text{d4},\delta(k)}T_{\text{in}}),\end{aligned} \tag{5.52}$$

with substituting all inputs that are not influenced by the online control input update. $\boldsymbol{d}_{\text{s}}(k)$ is bounded as $\boldsymbol{d}(k)$, $\boldsymbol{P}_t(k)$, T_{∞} and T_{in} are all bounded.

Defining a new state vector

$$\check{\boldsymbol{T}}(k) = \left[\boldsymbol{T}^T(k), \boldsymbol{e}^T(k)\right]^T, \tag{5.53}$$

a new system combining the system and the observer error dynamic is given by

$$\begin{aligned}\underbrace{\begin{bmatrix}\boldsymbol{T}(k+1)\\ \boldsymbol{e}(k+1)\end{bmatrix}}_{\check{\boldsymbol{T}}(k+1)} =& \underbrace{\begin{bmatrix}\boldsymbol{A}_{\text{d},\delta(k)} - \boldsymbol{B}_{\text{d1},\delta(k)}\boldsymbol{F}_{\delta(k)} & \boldsymbol{B}_{\text{d1},\delta(k)}\boldsymbol{F}_{\delta(k)}\\ 0 & \boldsymbol{A}_{\text{d},\delta(k)} - \boldsymbol{L}_{\delta(k)}\boldsymbol{C}_{\text{o}}\boldsymbol{A}_{\text{d},\delta(k)}\end{bmatrix}}_{\check{\boldsymbol{A}}_{\delta(k)}}\underbrace{\begin{bmatrix}\boldsymbol{T}(k)\\ \boldsymbol{e}(k)\end{bmatrix}}_{\check{\boldsymbol{T}}(k)} \\ &+ \underbrace{\begin{bmatrix}\boldsymbol{I} & 0\\ 0 & \boldsymbol{I} - \boldsymbol{L}_{\delta(k)}\boldsymbol{C}_{\text{o}}\end{bmatrix}}_{\check{\boldsymbol{H}}_{\delta(k)}}\underbrace{\begin{bmatrix}\boldsymbol{d}_s(k)\\ \boldsymbol{d}(k)\end{bmatrix}}_{\check{\boldsymbol{d}}(k)},\end{aligned} \tag{5.54}$$

Therefore, the new system is given by

$$\check{\boldsymbol{T}}(k+1) = \check{\boldsymbol{A}}_{\delta(k)}\check{\boldsymbol{T}}(k) + \check{\boldsymbol{H}}_{\delta(k)}\check{\boldsymbol{d}}(k) \tag{5.55}$$

Define a $\mathscr{L}_2$-gain between the unknown input vector $\check{\boldsymbol{d}}$ and the state vector $\check{\boldsymbol{T}}$

$$\gamma_{\text{s}} = \sup_{\left\|\check{\boldsymbol{d}}(k)\right\|_2 \neq 0} \frac{\left\|\check{\boldsymbol{T}}(k)\right\|_2}{\left\|\check{\boldsymbol{d}}(k)\right\|_2}. \tag{5.56}$$

The $\mathscr{L}_2$-gain stability of the whole system can be tested by the following theorem.

Theorem 5.2. *The switched system (5.11) with the proposed observer (5.15) and controller (5.49) is $\mathscr{L}_2$ stable, if there exits a scalar $\gamma_s > 0$ and a matrix $\mathscr{P}_s$ symmetric and positive definite, $\forall \delta \in \{1, ..., M_a\}$ such that*

$$\begin{bmatrix} \mathscr{P}_s - \breve{\boldsymbol{A}}_\delta^T \mathscr{P}_s \breve{\boldsymbol{A}}_\delta - \boldsymbol{I} & -\breve{\boldsymbol{A}}_\delta^T \mathscr{P}_s \breve{\boldsymbol{H}}_\delta \\ * & \gamma_s^2 \boldsymbol{I} - \breve{\boldsymbol{H}}_\delta^T \mathscr{P}_s \breve{\boldsymbol{H}}_\delta \end{bmatrix} > 0. \tag{5.57}$$

Proof. Define a Lyapunov function candidate for the switched system (5.54)

$$V(\breve{\boldsymbol{T}}(k)) = \breve{\boldsymbol{T}}^T(k) \mathscr{P}_s \breve{\boldsymbol{T}}(k), \tag{5.58}$$

where $\mathscr{P}_s$ is symmetric and positive definite. We suppose the follow inequality is satisfied

$$V(k+1) - V(k) + \breve{\boldsymbol{T}}^T(k)\breve{\boldsymbol{T}}(k) - \gamma_s^2 \breve{\boldsymbol{d}}^T(k)\breve{\boldsymbol{d}}(k) < 0 \tag{5.59}$$

for all $\breve{\boldsymbol{T}}(k)$ and $\breve{\boldsymbol{d}}(k)$. Substituting (5.55) into (5.58), one has

$$V(k+1) = \Big(\breve{\boldsymbol{A}}_{\delta(k)}\bar{\boldsymbol{T}}(k) + \breve{\boldsymbol{H}}_{\delta(k)}(k)\breve{\boldsymbol{d}}(k)\Big)^T \mathscr{P}_s \Big(\breve{\boldsymbol{A}}_{\delta(k)}\bar{\boldsymbol{T}}(k) + \breve{\boldsymbol{H}}_{\delta(k)}(k)\breve{\boldsymbol{d}}(k)\Big). \tag{5.60}$$

Therefore, inequality (5.59) can be written as

$$\begin{bmatrix} \breve{\boldsymbol{T}}(k) \\ \breve{\boldsymbol{d}}(k) \end{bmatrix}^T \begin{bmatrix} \breve{\boldsymbol{A}}_{\delta(k)}^T \mathscr{P}_s \breve{\boldsymbol{A}}_{\delta(k)} + \boldsymbol{I} - \mathscr{P}_s & \breve{\boldsymbol{A}}_{\delta(k)}^T \mathscr{P}_s \breve{\boldsymbol{H}}_{\delta(k)} \\ * & \breve{\boldsymbol{H}}_{\delta(k)}^T \mathscr{P}_s \breve{\boldsymbol{H}}_{\delta(k)} - \gamma_s^2 \boldsymbol{I} \end{bmatrix} \begin{bmatrix} \breve{\boldsymbol{T}}(k) \\ \breve{\boldsymbol{d}}(k) \end{bmatrix} < 0, \tag{5.61}$$

which is satisfied if

$$\begin{bmatrix} \mathscr{P}_s - \breve{\boldsymbol{A}}_\delta^T \mathscr{P}_s \breve{\boldsymbol{A}}_\delta - \boldsymbol{I} & -\breve{\boldsymbol{A}}_\delta^T \mathscr{P}_s \breve{\boldsymbol{H}}_\delta \\ * & \gamma_s^2 \boldsymbol{I} - \breve{\boldsymbol{H}}_\delta^T \mathscr{P}_s \breve{\boldsymbol{H}}_\delta \end{bmatrix} > 0. \tag{5.62}$$

is fulfilled for all $\delta \in \{1, ..., M_a\}$. This completes the proof. □

Remark 5.3. *In the physical system, the supply voltage and frequency, which is the input of the system, and the temperature of the cores, which is the state of the system, both have constraints. Here the constraints are not considered in the MPC controller design. The computational complexity is the reason why the constraints are not considered, as such an online optimal control policy is not implementable [LK99, LK01].*

5.3 Example

Here the 3D package multi-core system described in Section 4.4 is employed to verify the proposed H_∞ observer and MPC controller. The MCP layout is given in Figure 4.14 and Figure 4.15. The operation condition and the types of tasks are defined equivalently as in Section 3.3.

Assume the fluid is chosen from a set of three different velocities

$$\mathbb{U}_{\text{in}} = \{u_{\text{in},1}, u_{\text{in},2}, u_{\text{in},3}\} \tag{5.63}$$

where $u_{\text{in},1} = 0.5\,\text{m/s}$, $u_{\text{in},2} = 0.75\,\text{m/s}$ and $u_{\text{in},3} = 1\,\text{m/s}$. The possible target power zone is also divided in three intervals

$$\begin{aligned}\mathbb{P}_{\text{t},1} &= \{P_{\text{t}} | P_{\text{t}} < 4\,\text{W}\}\\ \mathbb{P}_{\text{t},2} &= \{P_{\text{t}} | 4\,\text{W} \leq P_{\text{t}} < 5\,\text{W}\}\\ \mathbb{P}_{\text{t},3} &= \{P_{\text{t}} | 5\,\text{W} \leq P_{\text{t}}\}\end{aligned} \tag{5.64}$$

while the possible area of the maximal temperature of dies T_{m} is divided in four intervals

$$\begin{aligned}\mathbb{T}_{\text{m},1} &= \{T_{\text{m}} | T_{\text{m}} < 313\,\text{K}\}\\ \mathbb{T}_{\text{m},2} &= \{T_{\text{m}} | 313\,\text{K} \leq T_{\text{m}} < 323\,\text{K}\}\\ \mathbb{T}_{\text{m},3} &= \{T_{\text{m}} | 323\,\text{K} \leq T_{\text{m}} < 328\,\text{K}\}\\ \mathbb{T}_{\text{m},4} &= \{T_{\text{m}} | 328\,\text{K} \leq T_{\text{m}}\}.\end{aligned} \tag{5.65}$$

Based on the MCP layout (Figure 4.14-4.15) and the possible fluid velocities (5.63) of the switched system (5.11) with three subsystems is deduced with the sampling period $t_s = 10\,\text{ms}$. The system output vector which contains the measurable states is set as the temperature of the 16 cores, the PPE and the L2 cache. The disturbance vector $\boldsymbol{d}(k)$ consists of the unknown power of the PPE, the L2 cache, the input/output controller and other heat generated by the interconnect bus. As the amount of tasks is usually randomly time-varying and as the unknown power is also varying with the type of tasks the disturbance vector can be assumed randomly time-varying and bounded which is simulated using a MATLAB function creating a random signal.

First the H_∞ observer is designed. By applying Theorem 5.1 using the MATLAB toolbox YALMIP [Löf04] and SeDuMi [Stu99], the LMI optimization gives the $\mathscr{L}_2$-gain $\gamma = 6.88$. Further the MPC controller is designed with two different cost functions. The first cost function is chosen as

$$J_1 = \sum_{i=1}^{3} \left[300\boldsymbol{\xi}_2^T(k+i)\boldsymbol{\xi}_2(k+i) + 2000\boldsymbol{P}_{\text{c}}^T(k+i-1)\boldsymbol{P}_{\text{c}}(k+i-1) \right] \tag{5.66}$$

and the second cost function is chosen as

$$J_2 = \sum_{i=1}^{3} \left[500\boldsymbol{\xi}_2^T(k+i)\boldsymbol{\xi}_2(k+i) + 2000\boldsymbol{P}_{\text{c}}^T(k+i-1)\boldsymbol{P}_{\text{c}}(k+i-1) \right]. \tag{5.67}$$

The two cost functions have the same weighting matrix of control input but a different weighting matrix of the regulated output. After computing the controller and observer gains, the stability is proved by applying Theorem 5.2. Under the cost function J_1

the $\mathscr{L}_2$-gain $\gamma_{s1} = 12$ is obtained, while under the cost function J_2 the $\mathscr{L}_2$-gain is also $\gamma_{s2} = 12$.

Simulation results for the core blocks

The simulation results under the cost function J_1 for the core block 7 (see Figure 4.14) are shown in Figure 5.2. The first plot in Figure 5.2 shows the resulting temperature of the die block with the deserved temperature region [313 K, 323 K] indicated by the dashed lines in the first plot Figure 5.2. The seconds plot shows the target power as well as the actual power resulting from the controller. It can be seen that the actual power is mostly reduced compared to the target power for the temperature balancing, i.e. $P_{c,7} > 0$. Therefore, for some other blocks the power is increased compared to the target power as can be seen Figure 5.3 for the core block 12. This shows that the tasks are moved to other cores based on the balancing controller.

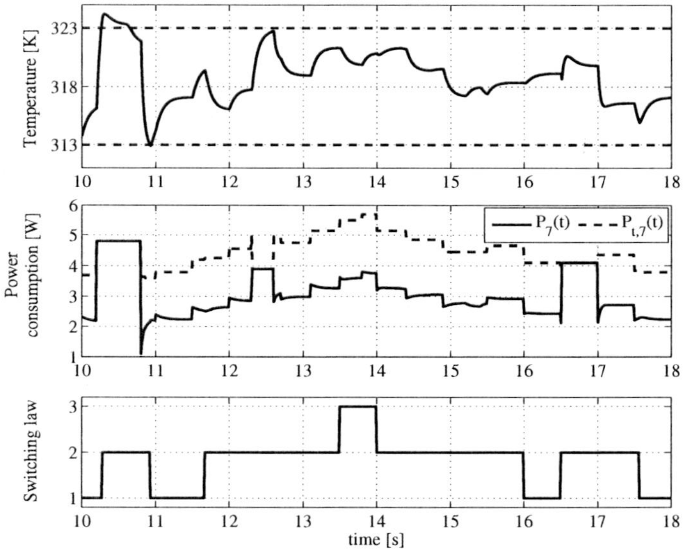

Figure 5.2: Simulation result for core block 7 under the cost function J_1

In the third plot the switching index is shown which represents the fluid velocity u_{in}. The switching index $\delta = 1$ indicates the fluid velocity $u_{in} = 0.5\,\mathrm{m/s}$, the index $\delta = 2$ indicates $u_{in} = 0.75\,\mathrm{m/s}$ and $\delta = 3$ represents $u_{in} = 1\,\mathrm{m/s}$. It can be clearly seen that the fluid velocity is increased from $u_{in} = 0.5\,\mathrm{m/s}$ to $u_{in} = 0.75\,\mathrm{m/s}$ when the temperature exceeds the boundary 323 K at the time instant $t = 10.3\,\mathrm{s}$. At this time instant the maximum temperature of the die is the temperature of block 7. This shows the functionality the velocity adjustment policy. In the remaining simulation time the velocity is adjusted due to the varying target power and temperature of all blocks which is not shown in detail for the other blocks.

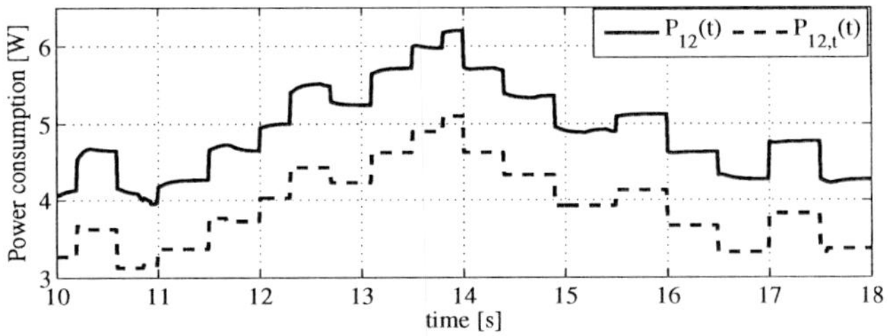

Figure 5.3: Power consumption for core block 12 under the cost function J_1

The third plot in Figure 5.2 also shows that during the time zone 10-18 s, the pump works for only 6.3 % of the time under the highest liquid velocity $u_{\text{in}} = 1\,\text{m/s}$ (i.e. $\delta = 3$). For 24.3 % of the time the controller works with $u_{\text{in}} = 0.5\,\text{m/s}$ (i.e. $\delta = 1$) while for the left 69.4 % it works with $u_{\text{in}} = 0.75\,\text{m/s}$ (i.e. $\delta = 2$). Therefore, the proposed method can let the pump work slower in most time, which leads to energy savings with respect to the pump energy.

Comparison of the simulation results for the core blocks

The comparison of the performance under the two different cost functions is shown in Figure 5.4 and 5.5. As shown in the two figures, cost function J_1 leads to lower 2-Norm of the control input while cost function J_2 results in lower temperature difference. This means under J_2 a better temperature balance can be achieved. However the power consumption needs to be adjusted more. For physical implementation, the aim is to reduce the temperature difference among the cores as small as possible. However, as shown in the simulation, a smaller temperature difference is accompanied by bigger power consumption adjustment, and causes more task adjustment among the cores.

In this example, due to the temperature balancing core 1, core 2, core 9, core 10 work at much higher speeds as these cores have better cooling condition, meanwhile, core 7, core 8 work at low speeds. This causes an unbalanced task assignment, which also affects the chip lifespan and working reliability. Besides, in an operation system, when the core's supply voltage and working frequency is adjusted, some tasks need to be moved from the lower frequency cores to the higher frequency cores. A higher power adjustment often causes more task movement, and this action results in additional time and energy.

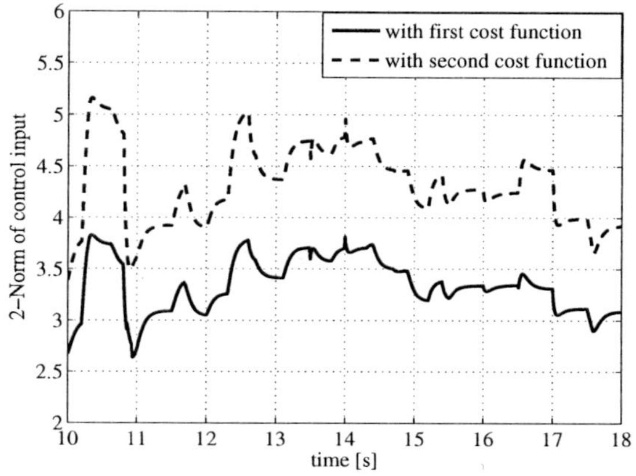

Figure 5.4: Comparison of the control input under the cost function J_1 and J_2

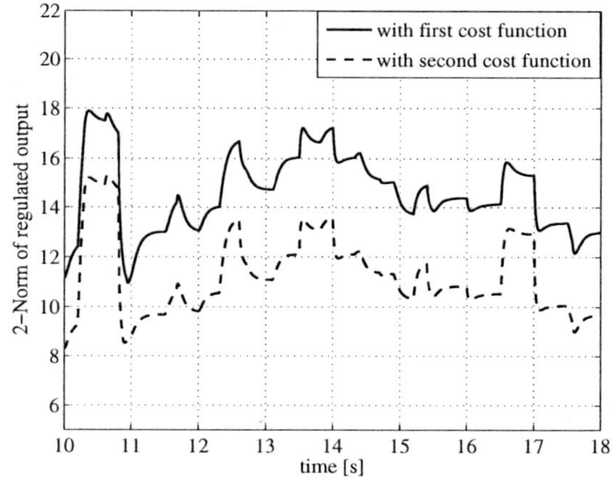

Figure 5.5: Comparison of the regulated output under cost the function J_1 and J_2

In summary, the pursuit of very small temperature difference among cores is not always practical in a physical system. To implement the proposed MPC controller in a 3D MCP system, the temperature balance, power management should be considered jointly.

The aim should be to balance the temperature, the power consumption and the task assignment.

Simulation results for the die blocks (non-core)

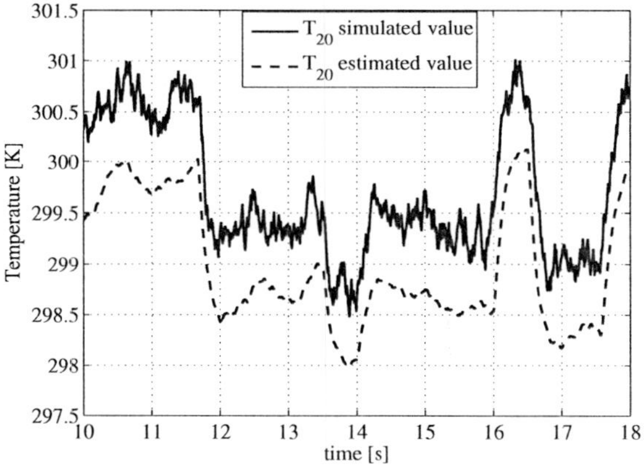

Figure 5.6: Real and estimated state of T_{20}

For the non-core die blocks only the state estimation is relevant, as this estimation is required for the control of the temperature of the core blocks. The temperature of those blocks is not controlled. Figure 5.6 shows the state estimation of the die block T_{20}, which is an I/O controller. For this block the power consumption is not known and thus acts as a disturbance on the state estimation. Therefore, the estimation error is comparingly large. However, as an H_∞ observer is applied the influence of the disturbance on the estimation can be bounded.

Simulation results for the fluid blocks

The estimated liquid temperature and the simulated liquid temperature is shown in Figure 5.7 - 5.8. The indices indicate the blocks according to Figure 4.15. The simulation time is 20 s whereas in the following figures only the time interval 10-18 s is shown.

Figure 5.8 shows the state estimation of T_{39} in the cooling layer 2 between the second layer and third layer of dies. This state and the state T_{40} have the biggest estimation error from the cooling layer states. The estimation error is only little affected by the disturbance. However, as it is a neighboring block of the I/O controller the estimation is strongly influenced by the estimation error of T_{20}. Additionally the disturbed estimations of the other blocks in the third die layer have some influence on the estimation of T_{39}.

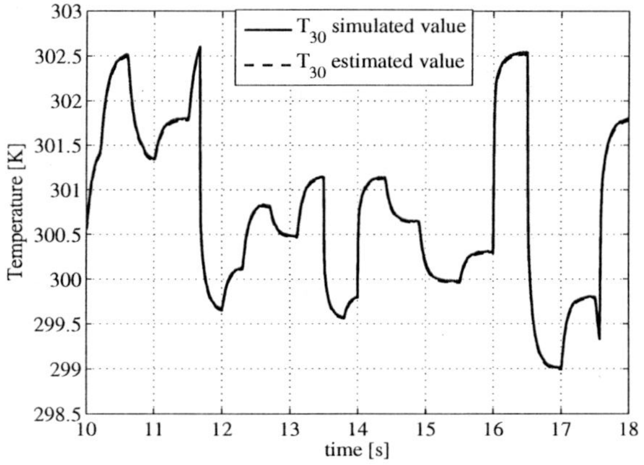

Figure 5.7: Real and estimated state of T_{30}

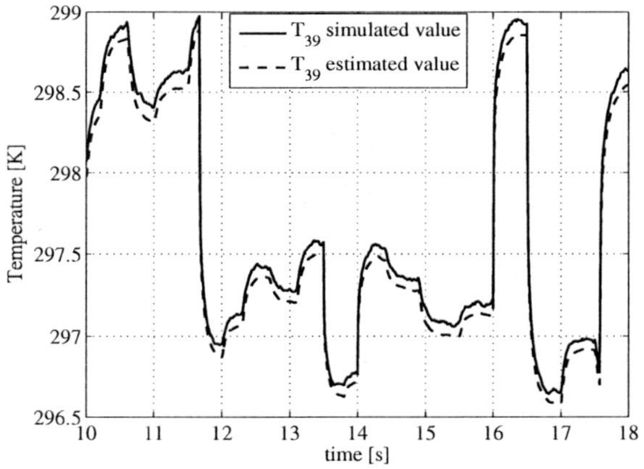

Figure 5.8: Real and estimated state of T_{39}

On the other hand the state estimation of T_{30} shown in Figure 5.7 is almost equivalent to the simulated value. As the shown state T_{30} is on the cooling layer 1 between the first layer and second layer of the dies and as the majority of the disturbance is caused by the power of parts in the third die layer those states are very little affected by the disturbance, see Figure 4.15. Thus the estimation error is negligible for the state estimations in the cooling layer 1.

5.4 Summary

The thermal and power management policy for 3D stacked package MCPs is designed in this chapter based on the model shown in the previous chapter. The management policy contains two steps. The first step is to control the liquid velocity based on the temperature of the dies and the real time workload. Under this policy, the system is described as a linear switched system. An MPC controller and an H_∞ observer is developed for the switched linear system in the second step. The simulation results under different conditions support the proposed control policy.

6 Summary and outlook

In this thesis, the modeling and control technologies for the thermal and power balancing/management of the multi core processor are developed. The thermal and power management is a technological challenge in MCP developing, as it has considerable influence on the chip lifespan and the operation reliability. Therefore, a feasible thermal and power management policy for the MCP system is a necessary and urgent task.

The heat transfer in the die is described as a 3D partial differential equation according to the Fourier's heat conduction law. However, the 3D PDE model is not suitable to be applied in the controller design as our target is to balance the temperature and manage the power among the cores. Hence, the 3D PDE is transformed to a group of 1D PDEs, which is applied for the control design. In this thesis, the 3D PDE heat transfer model is solved by the eigenvalue and eigenfunction system according to the Sturm-Liouville theory. The 3D to 1D transformation is applied based on the analytical solution.

A PDE based optimal controller is designed to manage and balance the temperature and the power consumption of the MCPs. A cost function is introduced which weights the difference of the temperature among the cores and the power consumption. The Riccati function is applied to optimize the cost function.

The advanced 3D stacked package chip technology is also considered in the thesis. The PDE based model is very complex as it contains the fluid dynamical model and heat transfer model. Therefore, an ODE model is proposed, which models the dies as an R-C network. Meanwhile the cooling channel is discretized based on the discrete block of the dies. The micro-channel liquid velocity and the convective heat transfer coefficient have a non-linear influence on the model. In view of this, the liquid velocity is chosen from a set of constant velocities which need to be defined in the control design. The definition of this set and the switching between the possible velocities is based on the real time tasks and the temperature of the dies. As the variation of the convective heat transfer coefficient is small, a simplification is realized by the taking an average value. By these specifications the non-linear terms can be eliminated and the system is modeled as a linear switched system where each subsystem refers to the model under one liquid velocity.

The control design for the thermal and power management is realized in two steps. In the first step a logic algorithm for determining the liquid velocity, i.e. the switching law is developed. In the second step a model predictive controller is designed to balance the temperature among the cores. Therefore a cost function weighting the difference of the temperature between the cores and the control input is introduced. Thereby the control input refers to the adjustment of the target power consumption of the cores. As not all necessary states can be assumed to be measurable a robust H_∞ observer is designed to estimate the states. The robust H_∞ observer is chosen to decrease the influence of unknown inputs like the unknown parts of the power consumption of the dies. Further a theorem is proposed to prove the stability of the controlled switched system under the separate control and observer design. Finally the results are verified by simulation.

For future work it is interesting to validate the modeling approaches of the 2D package chip and 3D package chip by practical experiments. In this context the exact identification of the parameters needs to be investigated. After the model validation the proposed control approaches can be implemented and verified. For the implementation of the proposed controllers in a physical MCP, the operating system task allocation policy should be considered. It is helpful to develop a method which can combine the proposed controllers and the operating system task allocation policies.

7 Summary in German

In der berühmten Rede 'There's Plenty of Room at the Bottom' (dt. Ganz unten ist eine Menge Platz oder Viel Spielraum nach unten) stellte Richard Feynman die Frage 'Warum können wir nicht die gesamten 24 Bände der Encyclopaedia Britannica auf dem Kopf einer Stecknadel speichern?'. Dies wird häufig als die Gründungsschrift der Nanotechnologie angesehen. Vom ersten Intel Prozessor 4004 mit 2000 Transistoren zum Intel Pentium 8400EE mit 2,3 Milliarden Transistoren aus dem Jahre 2010 hat der Integrationsgrad um das hunderttausend-fache zugenommen wobei Größe des Dies abgenommen hat. Nach dem Mooreschen Gesetz verdoppelt sich die Komplexität integrierter Schaltkreise mit minimalen Komponentenkosten regelmäßig alle zwei Jahre. Jedoch hat die Leistungssteigerung der Chips abgenommen, da Transistoren nicht grenzenlos kleiner werden können. Unterdessen erfordert die ständige Zunahme an Daten eine höhere Prozessorleistungen. Dies führte zur Entwicklung des Mehrkernprozessors, der mehrere Rechenkerne enthält. Jeder Kern besteht aus einer Recheneinheit und einem L1 Cache-Speicher. Manche Mehrkernprozessoren enthalten zusätzlich noch einen L2 Cache-Speicher. All diese Komponenten sind zur Kommunikation über ein schnelles Bussystem, dem sogenannten Element Interconnect Bus, miteinander verbunden.

Der Mehrkernprozessor weist die Vorteile der parallelen Ausführung mehrerer Task, einer höheren Prozessorgeschwindigkeit und einer besseren Energieeffizienz auf. Aufgrund der sich ergebenden deultichen Leistungssteigerung von Mehrkernprozessoren gegenüber Einkernprozessoren sind Mehrkernprozessoren der neue Trend in der Prozessorentwicklung.

Vom ersten Mehrkernprozessor Power4 mit zwei Kernen, der 1999 von IBM entwickelt wurde, bis zum TilePro 64 Prozessor mit 64 Kernen wurden signifikante Fortschritte im Bereich der Mehrkernprozessortechnology erzielt. Mehrere Entwicklungen, zum Beispiel der IBM cell oder der Sun Niagara, werden bereits erfolgreich in kommerziellen Anwendungen eingesetzt. Diese Mehrkernprozessoren sind in einer 2D Architektur aufgebaut, das heißt in dem Chip liegt nur eine Die-Schicht vor. Beim 2D-Packaging verbraucht das Netzwerk innerhalb des Dies viel Energie und es gibt eine hohe Wärmeentwicklung, was die Entwicklung der Mehrkernprozessortechnology einschränkt. Daher ist die 3D-Integration von großem Interesse. Beim 3D-Packaging enthält der Chip mindestens zwei Die-Schichten, welche vertikal verbunden sind durch Silikon-Durchkontaktierung (engl. through-silicon via, TSV) . Mit Hilfe der TSV-Technologie können Daten schneller zwischen den Kernen übertragen werden bei geringerem Energieverbrauch und geringerer Wärmeentwicklung. Um die Temperatur innerhalb der Chips zu reduzieren, wird ein

Mikrokanal-Kühlsystem zwischen den Schichten eingesetzt.

Sowohl in der 2D- wie auch in der 3D-Technologie ist das Temperatur- und Leistungsmanagement von zunehmender Bedeutung. In der Entwicklung von Mehrkernprozessoren stellt das Temperatur- und Leistungsmanagement eine zentrale technologische Herausforderung dar, da es einen signifikanten Einfluss auf die Betriebssicherheit und die Chiplebensdauer hat.

Motiviert durch die ausgeführten Aspekte hat diese Dissertation das Ziel, ein geeignetes Temperatur- und Leistungsmanagement zu entwickeln, um das thermische Verhalten und den Energieverbrauch von Mehrkernprozessoren zu regeln. Dadurch ergeben sich die folgenden Aufgabenstellungen. In einem ersten Schritt wird das thermische Verhalten eines 2D-Packaging Mehrkernprozessors modelliert. Anschließend wird dafür eine optimale Regelungsstrategie für das Temperatur- und Leistungsmanagement entwickelt. In einem weiteren Schritt werden Mehrkernprozessoren mit 3D-Technologie und einem Flüssigkeitskühlsystem untersucht. Nach der Modellierung des thermischen Verhaltens wird eine entsprechende Regelungsstrategie entworfen.

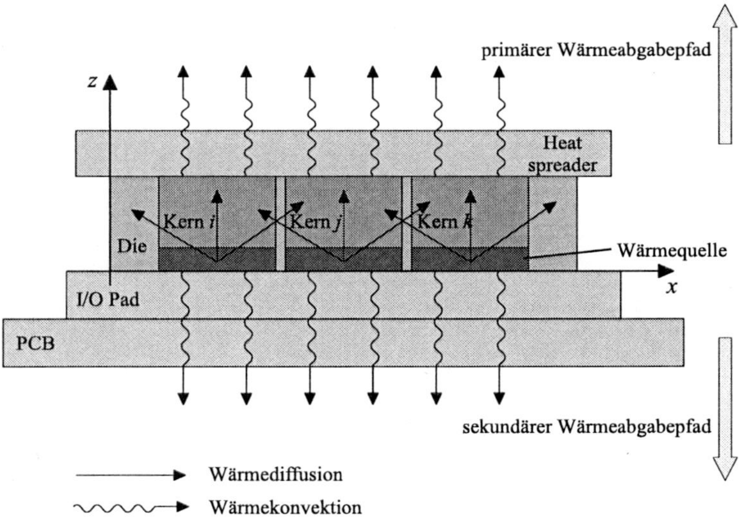

Figure 7.1: Die Querschnittsansicht der Mehrkernprozessor

In Kapitel 2 wird die Wärmekonduktion im Die sowie die Wärmekonvektion zwischen dem Die und der Umgebung analysiert, siehe Figure 7.1. Die Temperatur des Kerns hängt von der Leistungsabgabe des Kerns ab, wobei sich die Leistungsabgabe aus der Versorgungsspannung und der Arbeitsfrequenz ergibt. Sowohl die Versorgungsspannung als auch die Arbeitsfrequenz können durch Dynamic Voltage und Frequency Scaling (DVFS) angepasst werden. Die Wärmeübertragung innerhalb des Dies erfolgt durch Konduktion und kann basierend auf dem Fourieschen Wärmeleitungsgesetz als partielle Differentialgleichung (engl. partial differential equation, PDE) modelliert werden. Der Wärmeaustausch zwischen dem Die und der Umgebung erfolgt durch Konvektion und kann mit Hilfe des Newton'schen Abkühlungsgesetz (engl. Newton's law of cooling) beschrieben werden. Dadurch ergeben sich Randbedingungen für die dreidimensionale partielle Differentialgleichung. Die Sturm-Liouville-Theorie wird anschließend angewandt, um die dreidimensionale partielle Differentialgleichung zu lösen und eine Temperaturverteilung über das Volumen zu erhalten. Dies bedeutet, dass basierend auf den Eigenwerten und Eigenfunktionen die partielle Differentialgleichung auf ein System gewöhnlicher Differentialgleichungen (engl. ordinary differential equation, ODE) unendlicher Ordnung abgebildet wird. Nach dem Lösen des ODE Systems ergibt sich nach der inversen Transformation die dreidimensionale Wärmeverteilung innerhalb des Dies. Kapitel 2 bildet die Grundlage für den Reglerentwurf des 2D-Package Systems.

Kapitel 3 behandelt den optimal Reglerentwurf, wofür das 3D PDE Modell in ein System eindimensionaler PDEs transformiert wird. Während in Kapitel 2 das dreidimensionale Gesamtsystem des Dies modelliert wird, ist es nun von Interesse die Temperatur und die Leistungsabgabe zwischen den Kernen zu balancieren. Da in jedem Kern ein individuelle Temperaturverteilung vorliegt, muss festgelegt werden, wie der Temperaturunterschied zwischen den Kernen definiert wird. Daher wird jeder Kern individuell betrachtet und die anderen Kerne wirken als externe Wärmequellen, zu denen ein Wärmeaustausch erfolgt. Jeder Kern wird dabei durch eine eindimensionale partielle Differentialgleichung beschrieben. Die Ortsdimensionen der partiellen Differentialgleichungen der Kerne verlaufen dabei parallel zueinander. Die anderen beiden Dimensionen werden dadurch eliminiert, indem die Durchschnittstemperatur in Bezug auf die Ebene betrachtet wird. Der Temperaturunterschied zweier Kerne ist somit ortsabhängig durch die Differenz der Durchschnittstemperaturen gegeben. Zur Gesamtmodellierung ist noch die Beschreibung des gegenseitige thermischen Einflusses zwischen den Kernen erforderlich. Nach dem Fourieschen Wärmeleitungsgesetz hängen die Wärmeflüsse an den Rändern jedes Kerns vom Temperaturgradient an dem Rand ab, welcher basierend auf der in Kapitel 2 erhaltenen Lösung der dreidimensionalen PDE bestimmt wird. Es wird angenommen, dass an jedem Kern mindestens ein thermischer Sensor angebracht ist. Um die Leistungsabgabe und den Temperaturunterschied zwischen den Kernen zu regeln wird im weiteren ein optimaler Regler entworfen. In Abhängigkeit der auszuführenden Tasks wird zuerst eine äquivalente initiale Leistungsabgabe für jeden Kern definiert. Diese wird durch den optimalen Regler angepasst um die Temperaturdifferenz zwischen den Kernen zu reduzieren unter Berücksichtigung der Leistungsabgabe der Kerne. Dafür wird eine quadratische Kostenfunktion definiert, die sowohl die Temperaturdifferenz zwischen den Kernen als auch die Abweichung der Leistungsabgabe zur initialen Leistungsabgabe gewichtet. Mit Hilfe der Riccati Gleichung wird das optimale Regelungsproblem gelöst. Der Ausgang des optimalen Reglers ist die Anpassung der initialen Leistungsabgabe. Basierend auf der DVFS Technologie wird die Versorgungsspannung und die Arbeitsfrequenz entsprechend angepasst für jeden Kern. Kapitel 3 wird durch eine Simulation abgeschlossen, die die Ergebnisse verifiziert.

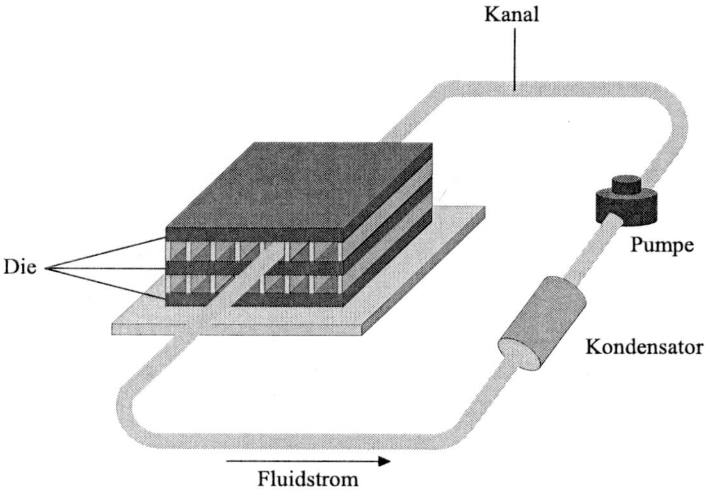

Figure 7.2: 3D-Package-Struktur

Die Modellierung des thermischen Verhaltens der fortgeschrittenen 3D-Packaging Chips wird in Kapitel 4 diskutiert. Wie bereits erwähnt wird ein Mikrokanal-Kühlsystem verwendet um die Temperatur zu reduzieren, siehe Figure 7.2. Das thermische Verhalten der Flüssigkeit des Kühlsystems hat dabei entscheidenden Einfluss auf die Temperatur der Dies. Außerdem tritt in Mikrokanälen das 'Thermal Wake' Phänomen auf, was die Beschreibung des Wärmeaustauschs komplex macht. Zur Beschreibung der Fluiddynamik der Flüssigkeit werden die Navier-Stokes Gleichungen verwendet. Durch zusätzliche Anwendung der Wärmeleitungsgleichung von Fluiden wird das dynamische thermische Modell hergeleitet, welches durch eine eindimensionale partielle Differentialgleichung beschrieben wird. Um dieses System zu lösen wird ein einfacher Algorithmus angewandt. Basierend auf der thermischen dynamischen Charakteristik des Fluids kann der Mikrokanal in Blöcke unterteilt werden, so dass das thermische Verhalten des Mikrokanals durch ein System gewöhnlicher Differentialgleichungen beschrieben werden kann. Genauso werden auch die Dies in Blöcke unterteilt wobei das thermische Verhalten eines Blocks durch eine Wärmekapazität und thermische Widerstände approximiert wird. Durch Zusammenfassung aller Blocks wird der gesamte Die durch ein R-C Netzwerk modelliert. Durch Kombination der Modelle des Dies und des Mikrokanals kann das gesamte thermische Verhalten beschrieben werden. Im weiteren werden der Einfluss von Geschwindigkeit und Temperatur der Kühlflüssigkeit auf die Performance des Kühlsystems diskutiert.

Die Regelungsstrategie für das in Kapitel 4 beschriebene 3D-Packaging System wird in Kapitel 5 entworfen. Es gibt zwei Ansätze um die Leistungsabgabe und Temperaturdifferenz zwischen den Kernen zu regeln. Ein Ansatz ist die Anpassung der Versorgungsspannung und der Arbeitsfrequenz der Kerne mittels DVFS Technologie. Ein weiterer Ansatz ist die Anpassung der Fluidgeschwindigkeit. Dabei wird berücksichtigt, dass mit steigender Geschwindigkeit der konvektive Wärmeübergangskoeffizient zunimmt. Außerdem führt eine höhere Geschwindigkeit zu größeren Temperaturunterschieden zwischen dem Fluid und den Dies, was einen positiven Einfluss auf die gewünschte Wärmekonvektion hat. Jedoch tritt die Geschwindigkeit nichtlinear in dem Modell auf, was zu hoher Komplexität sowohl für den Reglerentwurf als auch für die online Anpassung von Versorgungsspannung und Arbeitsfrequenz führt. Um die Komplexität zu reduzieren, wird der Regler in einem zweistufigen Verfahren entworfen. Zuerst wird die Fluidgeschwindigkeit durch einen einfachen logischen Algorithmus bestimmt. Dabei wird die Menge der Tasks und die größte auftretende Temperatur der Dies berücksichtigt. Die Fluidgeschwindigkeit ist somit für die zweite Stufe der Regelung bereits festgelegt. Basierend auf der Geschwindigkeitsanpassungsstrategie kann für jede mögliche Fluidgeschwindigkeit das System durch ein lineares Modell beschrieben werden. Durch die Definition einer begrenzten Anzahl möglicher Fluidgeschwindigkeiten wird das Gesamtsystem somit durch ein geschaltetes System modelliert, wobei jedes Subsystem die Dynamik unter einer bestimmten Fluidgeschwindigkeit beschreibt. Durch den Algorithmus zur Bestimmung der Fluidgeschwindigkeit ist somit auch das Schaltgesetz gegeben und somit bekannt. Als Systemausgang wird der Temperaturvektor der Blöcke, die die Temperatur der Kerne abbilden, und weiterer Blöcke (z.B. L2 Cache) gewählt. Die Auswahl der weiteren Blöcke geschieht in Abhängigkeit des Die-Layouts, um die Schätzung der nicht gemessenen Zustände mit ausreichender Genauigkeit sicherzustellen. Zu den nicht gemessenen Zuständen gehören insbesondere die Fluidtemperaturen der Blöcke, die durch einen geschalteten H_∞ Beobachter geschätzt werden. Zur Regelung basierend auf den geschätzten Zuständen wird schließlich ein geschalteter modellprädiktiver Regler (engl. Model Predictive Control, MPC) entworfen, um die Leistungsabgabe und Temperatur zwischen den Kernen zu balancieren. Das entwickelte Modell und die vorgestellte Regelungsstrategie mit einem geschalteten modellprädiktiven Regler sowie einem H_∞ Beobachter werden abschließend durch Simulation verifiziert. Dafür wird der 8 Kernprozessor IBM Cell und das von IBM Research in Zusammenarbeit mit EPFL und ETHZ entwickelte Mikrokanal-Kühlsystem als Beispiel genutzt. Die Simulationen zeigen die Eignung der Methodik zum Temperatur- und Leistungsmanagement.

A Mathematical Background

A.1 Gauss Theorem

Define Ω is a subset of $\mathbb{R}^r$ (in the case of this work, $r = 3$ and Ω represents a volume in 3D space), which is compact and has a piecewise smooth boundary S. If f is a continuously differentiable vector field defined on a neighborhood of Ω, according to the Divergence Theorem of Gauss shown in [SLS09, Ch. 6], one has

$$\iiint_\Omega \nabla f d\Omega = \oiint_S (f \cdot \boldsymbol{n}) dS \tag{A.1}$$

and equivalently

$$\iiint_\Omega \nabla^2 f d\Omega = \oiint_S \frac{\partial f}{\partial \boldsymbol{n}} dS. \tag{A.2}$$

A.2 Sturm-Liouville eigenvalue system

In the PDE equation (2.10) with the boundary conditions (2.13) - (2.16), the eigenvalue and eigenfunction problem can be solved by the Sturm-Liouville theory. A regular Sturm-Liouville problem has the form

$$\begin{aligned}\big(p(x)y'(x)\big)y'(x) + \big(q(x) + \lambda^2 r(x)\big)y(x) = 0,\ a < x < b \\ c_1 y(a) + c_2 y'(a) = 0 \\ d_1 y(b) + d_2 y'(b) = 0\end{aligned} \tag{A.3}$$

where $p(x)$, $q(x)$ and $r(x)$ are specific function and λ is a parameter. $(c_1, c_2) \neq (0, 0)$ and $(d_1, d_2) \neq (0, 0)$. $p(x)$, $p'(x)$, $q(x)$ and $r(x)$ are continuous on $[a, b]$. $p(x)$ and $r(x)$ are positive on $[a, b]$. A nonzero solution of $y(x)$ is called an eigenfunction while the corresponding value of λ is eigenvalue. The eigenvalues and eigenfunctions have the following characteristics,

1. There exists infinite number of eigenfunctions and eigenvalues, and

$$0 \leq \lambda_0 < \lambda_1 < \lambda_2 ... \tag{A.4}$$

2. Every pair of eigenfunctions $(y_m(x), y_n(x))$ are orthogonal with respect to the weight function $r(x)$, i.e.

$$\int_a^b y_m(x)y_n(x)r(x)dx = 0. \tag{A.5}$$

According to the theory of separation of variables [Pol01, Sup. B], a 3D PDE system shown in (2.10) with boundary conditions (2.12) can be sperated to x, y, z and t, four independent parts, which is

$$T(x,y,z,t) = T_x(x)T_y(y)T_z(z)T_t(t) \tag{A.6}$$

The eigenvalue system shown in (2.17) with boundary condition (2.18) is the eigenvalue system of the 3D system, which can also be separated in x-, y- and z-direction independently. In this case, $p(x)$, $q(x)$ and $r(x)$ are equal to 1, c_1, d_1 is related to h_k while c_2 and d_2 are equal to K.

A.3 Verification of the orthogonality of the eigenfunctions

Here we define the scalar product of two 3D functions as

$$(f_1, f_2) \equiv \iiint_\Omega f_1(x,y,z)f_2(x,y,z)dxdydz \tag{A.7}$$

Suppose the a pair of eigenfunctions are $\phi_{a_1b_1c_1}(x,y,z)$ and $\phi_{a_2b_2c_2}(x,y,z)$, where $a_1b_1c_1$ and $a_2b_2c_2$ are two indices of the eigenvalues and eigenfunctions satisfy the equation

$$\nabla^2\phi_{abc}(x,y,z) + \lambda_{abc}^2\phi_{abc}(z,y,z) = 0, \tag{A.8}$$

with boundary condition

$$K\frac{\partial\phi_{abc}(x,y,z)}{\partial \boldsymbol{n}} + h_k\phi_{abc}(x,y,z) = 0. \tag{A.9}$$

We can get

$$\begin{aligned}&(\lambda_{a_1b_1c_1}^2 - \lambda_{a_2b_2c_2}^2)\iiint_\Omega \phi_{a_1b_1c_1}(x,y,z)\phi_{a_2b_2c_2}(x,y,z)dxdydz\\ =&\iiint_\Omega [\nabla^2\phi_{a_1b_1c_1}(x,y,z) - \nabla^2\phi_{a_2b_2c_2}(x,y,z)]dxdydz.\end{aligned} \tag{A.10}$$

Then, according to the Gauss theorem A.1, one has

$$\begin{aligned}&\iiint_\Omega \phi_{a_1b_1c_1}(x,y,z)\phi_{a_2b_2c_2}(x,y,z)dxdydz\\ =&\frac{1}{(\lambda_{a_1b_1c_1}^2 - \lambda_{a_2b_2c_2}^2)}\sum_{k=1}^{6}\iint_{S_k}[\phi_{a_1b_1c_1}\frac{\partial\phi_{a_2b_2c_2}(x,y,z)}{\partial \boldsymbol{n}_l} - \phi_{a_2b_2c_2}\frac{\partial\phi_{a_1b_1c_1}(x,y,z)}{\partial \boldsymbol{n}_k}]ds_k.\end{aligned} \tag{A.11}$$

According to (A.9), we can get that if $a_1b_1c_1 \neq a_2b_2c_2$, one has

$$\iiint_{\Omega} \phi_{a_1b_1c_1}(x,y,z)\phi_{a_2b_2c_2}(x,y,z)dxdydz = 0. \tag{A.12}$$

If $a_1b_1c_1 = a_2b_2c_2$, then applied the L'Hôspital's rule on (A.11), we have

$$\begin{aligned}
&\iiint_{\Omega} \phi_{a_1b_1c_1}(x,y,z)\phi_p(x,y,z)dxdydz \\
=&\frac{1}{2\lambda_{a_1b_1c_1}} \sum_{k=1}^{6} \iint_{S_k} \begin{vmatrix} (\frac{\partial\phi(x,y,z)}{\partial\lambda})_{\lambda=\lambda_{a_1b_1c_1}} & (\frac{\partial^2\phi(x,y,z)}{\partial n_k \partial\lambda})_{\lambda=\lambda_{a_1b_1c_1}} \\ \phi_{a_1b_1c_1}(x,y,z) & \frac{\partial\phi_{a_1b_1c_1}(x,y,z)}{\partial n_k} \end{vmatrix} ds_k \\
=&\frac{1}{G_{a_1b_1c_1}} \neq 0,
\end{aligned} \tag{A.13}$$

which shows that each possible pair of eigenfunctions are orthogonal[CZ95, App. A].

A.4 L'Hôspital's rule

According to [Tay52], for two functions $f(x)$ and $g(x)$, which are differentiable on an open interval $\boldsymbol{I}$, and $a \in \boldsymbol{I}$. If

$$\lim_{x\to a} f(x) = \lim_{x\to a} g(x) = 0 \quad \text{or} \quad \pm\infty, \tag{A.14}$$

$\frac{f'(x)}{g'(x)}$ exits and $g'(x) \neq 0$, then

$$\lim_{x\to a} \frac{f(x)}{g(x)} = \lim_{x\to a} \frac{f'(x)}{g'(x)}. \tag{A.15}$$

A.5 Newton-Raphson method

The aim of applying the Newton-Raphson method [MÖ94] is to solve the equation

$$\frac{K^2\lambda_{zc}^2 - h_1h_2}{K\lambda_{zc}(h_1+h_2)} = \cot(\lambda_{zc}L_z). \tag{A.16}$$

The curves of the left hand side and right hand side are shown in Figure A.1. As the period of the right hand side is equal to π/z_1, the equation has one root in each period. Apply the Newton-Raphson method, with the initial guess of each root being $\pi c/z_1 + 0.4\pi/z_1$, and the interactive is given by

$$\lambda_{zc_{n+1}} = \lambda_{zc_n} - \frac{f(\lambda_{zc_n})}{f'(\lambda_{zc_n})}, \tag{A.17}$$

where

$$f(\lambda_{zc_n}) = \frac{K^2\lambda_{zc}^2 - h_1 h_2}{K\lambda_{zc}(h_1 + h_2)} - \cot(\lambda_{zc} z_1). \tag{A.18}$$

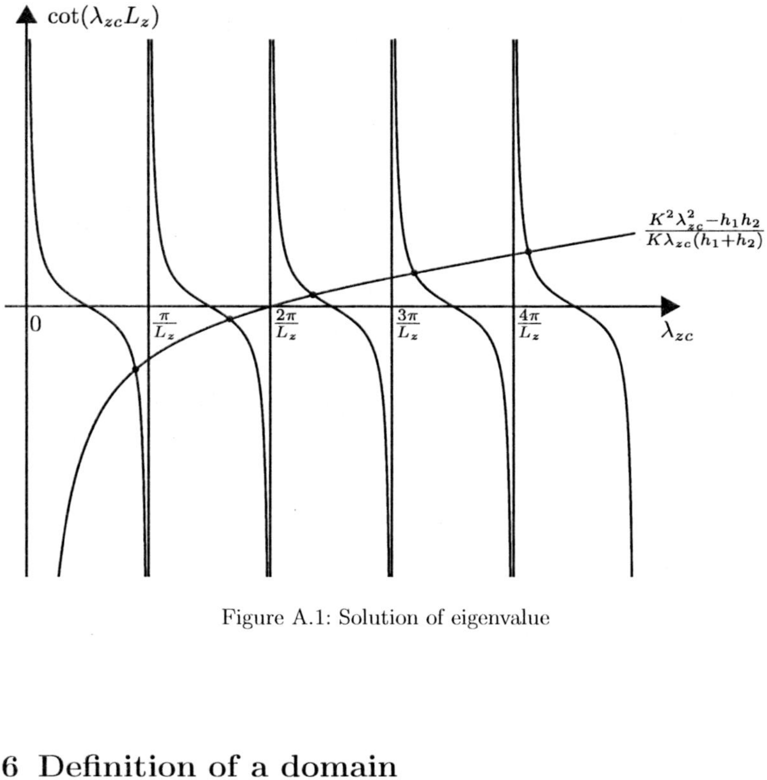

Figure A.1: Solution of eigenvalue

A.6 Definition of a domain

According to the physical fact, temperature variation is continuous and the heat conduction process is also continuous in the system shown in Chapter 2. Based on the semigroup theory in [CZ95, Sec. 2.1], the operator $\boldsymbol{A}$ and the domain of $\boldsymbol{A}$ are defined as follows.

$$
\begin{aligned}
\boldsymbol{A}T_{\mathrm{a},i}(z,t) =& \frac{d^2T_{\mathrm{a},i}(z,t)}{dz^2} \\
\boldsymbol{D}(\boldsymbol{A}) =& \Big\{ T_{\mathrm{a},i}(z,t) \in \mathscr{L}_2(0,T_{\mathrm{am}}) \ \Big| \ T_{\mathrm{a},i}(z,t), \frac{dT_{\mathrm{a},i}(z,t)}{dz} \text{ are absolutely continuous} \\
& \frac{d^2T_{\mathrm{a},i}(z,t)}{dz^2} \in \mathscr{L}_2(0,T_{\mathrm{am}}) \text{ and} \\
& \frac{dT_{\mathrm{a},i}(L_z,t)}{dz} = -h_1T_{\mathrm{a},i}(L_z,t).\ \frac{dT_{\mathrm{a},i}(0,t)}{dz} = h_2T_{\mathrm{a},i}(0,t) \Big\}
\end{aligned} \tag{A.19}
$$

In the set $\mathscr{L}_2(0,T_{\mathrm{am}})$, T_{am} is a constant defined at least as big as the maximum over-temperature of the cores. As in the MCP system, the power is bounded and the system is not adiabatic, there exists an upper bound of the over-temperature.

A.7 Solve the Riccati equation (3.44)

As described in Chapter 2, the PDE equation can be approached by a group of ODEs. The number of the ODE equation depends on the eigenvalue and eigenfunction system and the requirement of computing accuracy. Therefore, the Riccati equation (3.44) can be solved by applied the ODEs to approach this equation.

The PDE model (2.10) with the boundary conditions (2.13)-(2.16) can be approached by a group of ODEs with the following form

$$
\frac{d\overline{T}_{abc}(t)}{dt} + \lambda_{abc}^2\frac{K}{\sigma\rho}\overline{T}_{abc}(t) = \frac{1}{\sigma\rho}\sum_{i=1}^{N}\overline{q}_{abc,i}(t) + \frac{K}{\sigma\rho}\sum_{k=1}^{6}\iint_{S_k}\frac{\phi_{abc}(s_k)}{K}T_{\infty}(s_k,t)ds_k. \tag{A.20}
$$

Likewise, the system model (3.30) can be approached by a group of ODEs with the form

$$
\dot{\overline{\boldsymbol{T}}}_{\mathrm{a},c}(t) = \boldsymbol{A}_{zc}\overline{\boldsymbol{T}}_{\mathrm{a},c}(t) + \sum_{l=1}^{L_{\mathrm{d}}}\boldsymbol{B}_{zc,l}\boldsymbol{P}(t-l\cdot t_{\mathrm{d}}) \tag{A.21}
$$

where, $\dot{\overline{\boldsymbol{T}}}_{\mathrm{a},c} \in \mathbb{R}^{N\times 1}$ is the integral transform of $\dot{\boldsymbol{T}}_{\mathrm{a},c}$ with the transform is defined as

$$
\overline{f}_{zc}(t) = \int_0^{L_z}\phi_{zc}(z)f(z,t)dz \tag{A.22}
$$

and

$$
\boldsymbol{A}_{zc} = \mathrm{diag}\underbrace{\left(\lambda_{zc}^2\frac{K}{\sigma\rho}, \lambda_{zc}^2\frac{K}{\sigma\rho}, \ldots, \lambda_{zc}^2\frac{K}{\sigma\rho}\right)}_{N}, \tag{A.23}
$$

$$
\boldsymbol{B}_{zc,l} = \mathrm{diag}\underbrace{\left(\frac{1}{\sigma\rho}\sum_{l=1}^{L_{\mathrm{d}}}\overline{\gamma}_{1l}(z), \frac{1}{\sigma\rho}\sum_{l=1}^{L_{\mathrm{d}}}\overline{\gamma}_{il}(z), \ldots, \frac{1}{\sigma\rho}\sum_{l=1}^{L_{\mathrm{d}}}\overline{\gamma}_{Nl}(z)\right)}_{N}, \tag{A.24}
$$

with $\overline{\gamma}_l(z)$ is the integral transform of $\gamma_l(z)$ defined by (A.22).

Assume the system is approached by n eigenvalues and eigenfunctions, i.e. $c \in [0, ..., n-1]$, then the system can be approached as

$$\begin{bmatrix} \dot{\overline{\boldsymbol{T}}}_{\mathrm{a},0}(t) \\ \vdots \\ \dot{\overline{\boldsymbol{T}}}_{\mathrm{a},n-1}(t) \end{bmatrix} = \begin{bmatrix} \boldsymbol{A}_{z0} & & \\ & \ddots & \\ & & \boldsymbol{A}_{z(n-1)} \end{bmatrix} \begin{bmatrix} \overline{\boldsymbol{T}}_{\mathrm{a},0}(t) \\ \vdots \\ \overline{\boldsymbol{T}}_{\mathrm{a},l-1}(t) \end{bmatrix} + \sum_{l=0}^{L_{\mathrm{d}}} \begin{bmatrix} \boldsymbol{B}_{z0,l} \\ \vdots \\ \boldsymbol{B}_{z(n-1),l} \end{bmatrix} \boldsymbol{P}(t - l \cdot t_{\mathrm{d}}), \tag{A.25}$$

with the measurement output and the regulated output are

$$\boldsymbol{\xi}_1(t) = C_{\mathrm{o}} \left[\dot{\overline{\boldsymbol{T}}}_{\mathrm{a},0}(t), ..., \dot{\overline{\boldsymbol{T}}}_{\mathrm{a},n-1}(t) \right]^T \tag{A.26}$$

$$\boldsymbol{\xi}_2(t) = C_{\mathrm{r}} \left[\dot{\overline{\boldsymbol{T}}}_{\mathrm{a},0}(t), ..., \dot{\overline{\boldsymbol{T}}}_{\mathrm{a},n-1}(t) \right]^T \tag{A.27}$$

where C_{o} and C_{r} can be obtained from the inverse transform as described in Chapter 2. Based on the approached system (A.25) and (A.26), the PDE Riccati equation (3.44) can be solved as an ODE Riccati equation.

A.8 Schur complement

For a symmetrical matrix $\boldsymbol{S}$, and

$$\boldsymbol{S} = \begin{bmatrix} \boldsymbol{S}_{11} & \boldsymbol{S}_{12} \\ \boldsymbol{S}_{12}^T & \boldsymbol{S}_{22} \end{bmatrix} \tag{A.28}$$

where $\boldsymbol{S}_{11} \in \mathbb{R}^{r \times r}$, the following inequations are equivalent,

1. $\boldsymbol{S} < 0$;
2. $\boldsymbol{S}_{11} < 0$, $\boldsymbol{S}_{22} - \boldsymbol{S}_{12}^T \boldsymbol{S}_{11}^{-1} \boldsymbol{S}_{12} < 0$;
3. $\boldsymbol{S}_{22} < 0$, $\boldsymbol{S}_{11} - \boldsymbol{S}_{12} \boldsymbol{S}_{22}^{-1} \boldsymbol{S}_{12}^T < 0$,

see [BEFB94, Sec. 2.1].

A.9 Matrix inversion Lemma

The Matrix inversion Lemma has a form [Dun44] that

$$(\boldsymbol{A} + \boldsymbol{USV})^{-1} = \boldsymbol{A}^{-1} - \boldsymbol{A}^{-1}\boldsymbol{US}(\boldsymbol{S} + \boldsymbol{SVA}^{-1}\boldsymbol{US})^{-1}\boldsymbol{SVA}^{-1}, \tag{A.29a}$$

$$(\boldsymbol{A} - \boldsymbol{US}^{-1}\boldsymbol{V})^{-1} = \boldsymbol{A}^{-1} + \boldsymbol{A}^{-1}\boldsymbol{U}(\boldsymbol{S} - \boldsymbol{VA}^{-1}\boldsymbol{U})^{-1}\boldsymbol{VA}^{-1}, \tag{A.29b}$$

where $\boldsymbol{A} \in \mathbb{R}^{n \times n}$, $\boldsymbol{U} \in \mathbb{R}^{n \times r}$, $\boldsymbol{S} \in \mathbb{R}^{r \times r}$ and $\boldsymbol{V} \in \mathbb{R}^{r \times n}$ are matrices.

B Fluid dynamical and thermal physical background

B.1 Introduction to fluid parameters

The following introduction to the three parameters is from [Geb71].

Reynolds number is a dimensionless number that gives a measure of the ratio of inertial forces to viscous forces and consequently quantifies, the ratio, i.e.

$$Re = \frac{\rho V D_{\mathrm{h}}}{\mu}, \tag{B.1}$$

where V [m/s] is the mean velocity, D_{h} [m] is the hydraulic diameter, and

$$D_{\mathrm{h}} = 2W_{\mathrm{ch}}H_{\mathrm{ch}}/(W_{\mathrm{ch}} + H_{\mathrm{ch}}). \tag{B.2}$$

Prandtl number is a dimensionless number, which is defined as the ratio of momentum diffusivity (kinematic viscosity) to thermal diffusivity, i.e.

$$Pr = \frac{C_{\mathrm{f}}\mu}{K_{\mathrm{f}}}. \tag{B.3}$$

The Prandtl number is about 7 at 20°C.

Nusselt number is the ratio of heat convection to heat conduction across the boundary in the heat transfer processor at a boundary with fluid, i.e.

$$Nu = \frac{h_{\mathrm{f}}L_{\mathrm{f}}}{K_{\mathrm{f}}} \tag{B.4}$$

where h_{f} is the convective heat transfer coefficient, L_{f} is the channel characteristic length, and K_{f} is the thermal conductivity of the fluid. In practical systems, Nu can be used to get h_{f}, and the possible methods to get Nu can be found in [SL78, HKG99, QM02].

B.2 Body forces and surface forces

According to the description in [SB01], for a fluid element, two types of forces act on it, which are called body force and surface force. Body forces (see Figure B.1) act throughout the body of the fluid element. These forces are non-contact forces and distributed over the entire mass or volume of the element. Gravity, inertial force and electromagnetic force are body forces. Body forces ($d\boldsymbol{F}_b$) which act on the fluid element is proportional to the volume of the element ($d\boldsymbol{V}$).

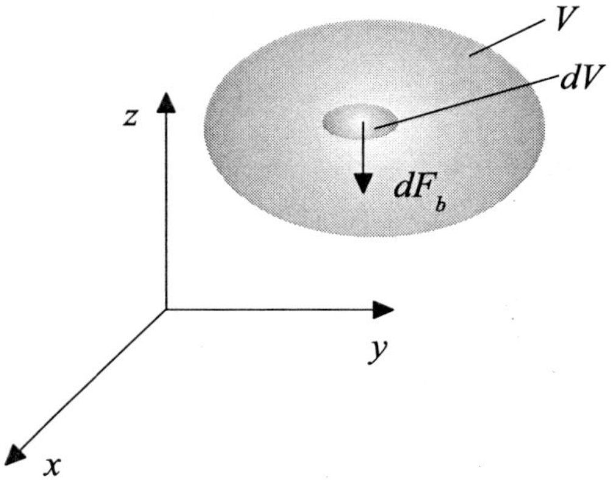

Figure B.1: Body force

Surface forces (see Figure B.2) are the forces which exert on the fluid element surface. These forces are contact forces and only appear on the surface of a fluid element. The normal force and shear force are surface force. The normal force is along the normal of an area while the shear force is along the plane of the area. Surface forces $d\boldsymbol{F}_s$ which act on element dS depend on the position of the volume and the area.

Stress tensor is a kind of surface force which contains both normal force and shear force. In Figure 4.5, τ_{xy}, τ_{xz}, τ_{yx}, τ_{yz}, τ_{zx} and τ_{zy} are shear force while τ_{xx}, τ_{yy} and τ_{zz} are normal force.

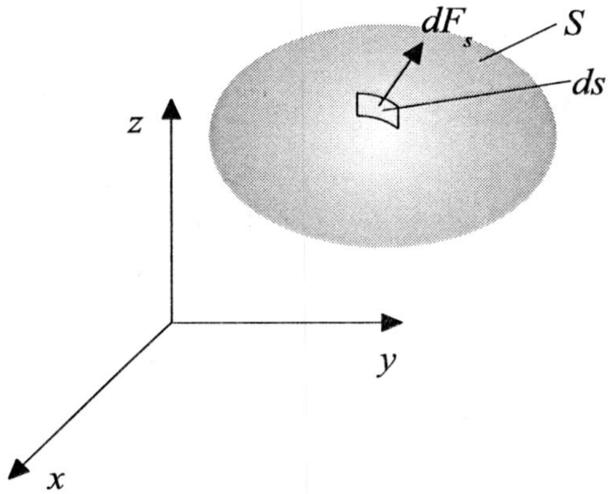

Figure B.2: Surface force

C SIMPLE alogirthm

The Semi-Implicit Method for Pressure Linked Equations (SIMPLE) algorithm [Pat80] can be used to solve the 3D pressure, velocity and temperature distribution. To employ this algorithm, first a grid should be designed and the system should be discretized based on the grid network. Taken a 1D static state PDE for example, the homogeneous equation of the momentum conservation is

$$\rho_f u \frac{\partial u}{\partial x} = \mu \frac{\partial^2 u}{\partial x^2} \tag{C.1}$$

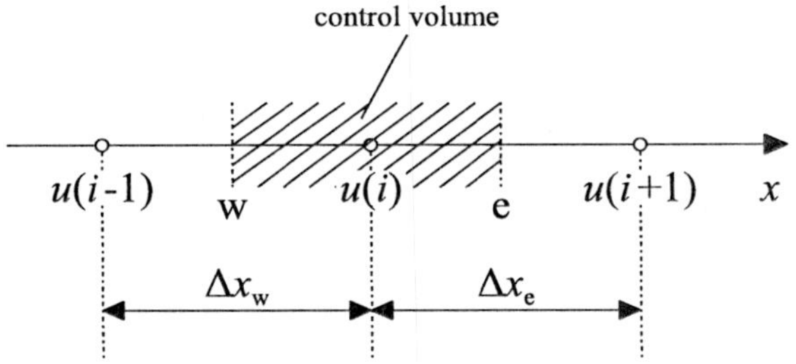

Figure C.1: The grid-point for a one-dimensional problem [Pat80]

Setting the grid as shown in Figure C.1 and integrating (C.1) from the point w to e, one has

$$(\rho_f u)_e u_e - (\rho_f u)_w u_w = \mu \left(\frac{du}{dx}\right)_e - \mu \left(\frac{du}{dx}\right)_w . \tag{C.2}$$

Considering u_e and u_w can be achieved based on the value of $u(i-1)$, $u(i)$ and $u(i+1)$, then one has

$$a_p u(i) = a_e u(i+1) + a_w u(i-1), \tag{C.3}$$

to get a_e and a_w. An interpolation algorithm needs to be employed. Meanwhile, in order to get a convergent result, a_e and a_w must be non-negative. The power-law scheme introduced in [Pat80] has been proved to be a feasible solution. Defining

$$F = \rho_f u, \qquad D = \frac{\mu}{\Delta x},$$

the power-law expressions for a_e is

$$a_e = D_e \max\left(0, \left(1 - \frac{0.1\,|F_e|}{D_e}\right)^5\right) + \max(0, -F_e), \tag{C.4}$$

where F_e and D_e are F and D in the positive x-direction. This solution is the upstream difference scheme, which considers the flow direction and the negative a_e and a_w can be avoided. The details can be found in [Pat80].

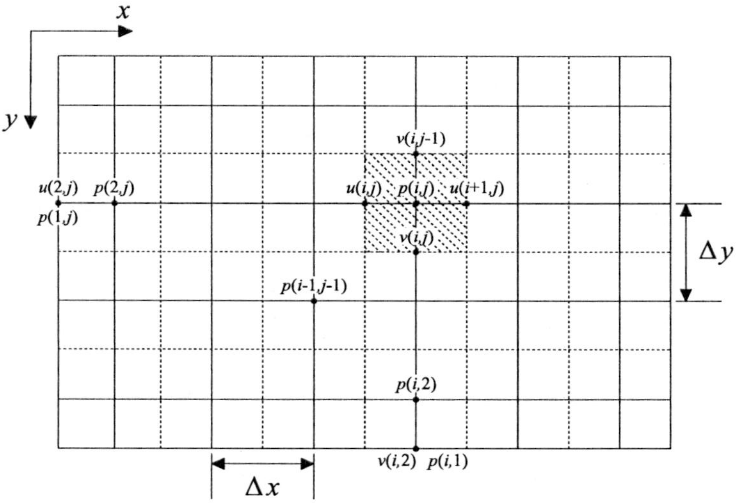

Figure C.2: The staggered grid network (top view)

A difficulty of the SIMPLE algorithm is that the pressure distribution in (4.7) is unknown. If we calculate the velocity and pressure distribution with the same grid, it will not contain the composition of forces on each point. The momentum equations are not effected by the pressure distribution, and this may lead to an undesirable solution. To avoid this problem, a staggered grid is a possible solution, which means that the pressure and velocity are not calculated at the same grid points as shown in Figure C.2 [Pat80]. Figure C.2 is the vertical view of the grid (in a fixed z-plane) and Figure C.3 is one calculation block of the 3D grid. In this grid network, the velocity u is calculated at the plane which is normal to the x-direction while v at the plane normal to the y-direction and w at the plane normal to the z-direction.

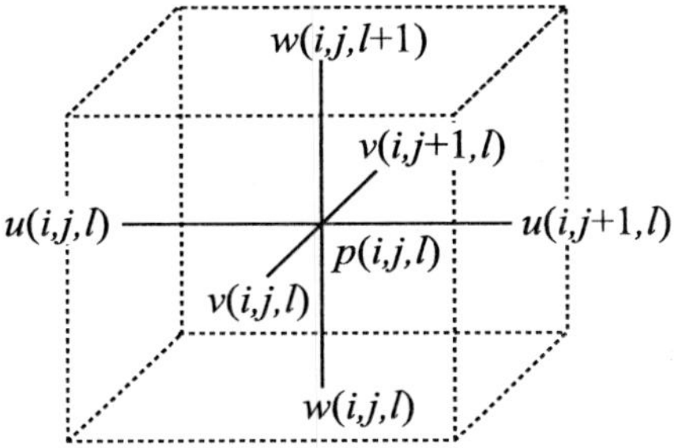

Figure C.3: The staggered grid network (3D)

Before giving out the algorithm, for a general distribution φ, define the w,e,s,n,b and t direction as shown in Figure C.4.

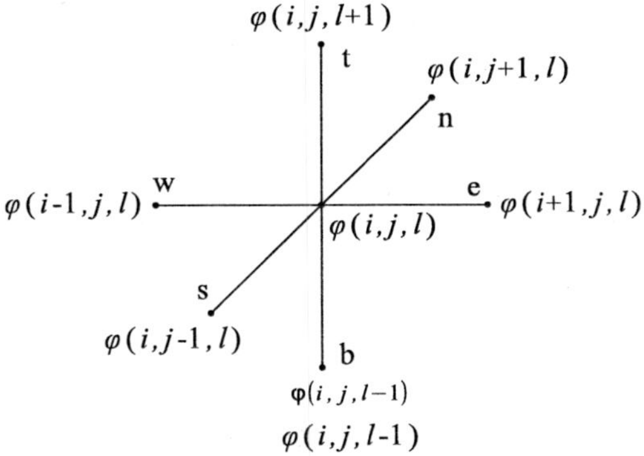

Figure C.4: The define of directions

The SIMPLE algorithm can be summarized in the following steps from [Pat80].

1. Guess a pressure field p^*.
2. Solve the momentum conservation equation and calculate u^*,v^*,w^*, the main algorithm is as follows

$$\begin{aligned} a_{\mathrm{p}}u^*(i,j,l,t) &= a_{\mathrm{w}}u^*(i-1,j,l,t)+a_{\mathrm{e}}u^*(i+1,j,l,t)+a_{\mathrm{s}}u^*(i,j-1,l,t) \\ &\quad +a_{\mathrm{n}}u^*(i,j+1,l,t)+a_{\mathrm{b}}u^*(i,j,l-1,t)+a_{\mathrm{t}}u^*(i,j,l+1,t)+b_{\mathrm{p}} \\ &\quad +S_{\mathrm{p}}(p^*(i-1,j,l,t)-p^*(i,j,l,t)), \end{aligned} \tag{C.5}$$

where * indates the guessed velocity and pressure. a_{w}, a_{e}, a_{s}, a_{n}, a_{b}, a_{t} are the coefficients which indicate the relationship between the estimated point and the adjacent points, with

$$\begin{aligned}
a_{\text{w}} &= D_{\text{w}} \max\left(0, (1 - \frac{0.1\,|F_{\text{w}}|}{D_{\text{w}}})^5\right) + \max(0, F_{\text{w}}) \\
a_{\text{e}} &= D_{\text{e}} \max\left(0, (1 - \frac{0.1\,|F_{\text{e}}|}{D_{\text{e}}})^5\right) + \max(0, -F_{\text{e}}) \\
a_{\text{s}} &= D_{\text{s}} \max\left(0, (1 - \frac{0.1\,|F_{\text{s}}|}{D_{\text{s}}})^5\right) + \max(0, F_{\text{s}}) \\
a_{\text{n}} &= D_{\text{n}} \max\left(0, (1 - \frac{0.1\,|F_{\text{n}}|}{D_{\text{n}}})^5\right) + \max(0, -F_{\text{n}}) \\
a_{\text{b}} &= D_{\text{b}} \max\left(0, (1 - \frac{0.1\,|F_{\text{b}}|}{D_{\text{b}}})^5\right) + \max(0, F_{\text{b}}) \\
a_{\text{t}} &= D_{\text{t}} \max\left(0, (1 - \frac{0.1\,|F_{\text{t}}|}{D_{\text{t}}})^5\right) + \max(0, -F_{\text{t}})
\end{aligned}$$

The water is considered as an ideal fluid, therefore, ι_x, ι_y and ι_z are assumed to be 0. b_{p} is the term of the influence on the state of the last calculation time instant.

$$b_{\text{p}} = \frac{a_{\text{p}}^0}{u(i,j,l,t-\Delta t)} \tag{C.6}$$

where

$$a_{\text{p}}^0 = \frac{\rho \Delta x \Delta y \Delta z}{\Delta t},$$

Δt is the calculation time interval. $S_{\text{p}}(p^*(i-1,j,l)$ describes the pressure force influence on $u(i,j,l,t)$ in the control volume and S_{p} is the area where the pressure difference has influence, therefore

$$S_{\text{p}} = \Delta y \Delta z, \tag{C.7}$$

and

$$a_{\text{p}} = a_{\text{w}} + a_{\text{e}} + a_{\text{s}} + a_{\text{n}} + a_{\text{b}} + a_{\text{t}} + a_{\text{p}}^0. \tag{C.8}$$

For the pressure p, only the two points in x-direction between $u(i,j,l,t)$ need to be considered, as u is the velocity in x-direction.

3. Solve the pressure correction equation, and achieve the pressure correction value. Suppose the pressure field has the following form

$$p(i,j,l,t) = p^*(i,j,l,t) + p'(i,j,l,t), \tag{C.9}$$

where p' is the pressure correction term. The pressure correction is based on the mass conservation equation (4.2). Discretizing this equation, one has

$$\begin{aligned}
&\rho\Big(u(i+1,j,l) - u(i,j,l)\Big)\Delta y \Delta z + \rho(v(i,j+1,l) \\
&-v(i,j,l))\Delta z \Delta x + \rho\Big(w(i,j,l+1) - w(i,j,l)\Big)\Delta x \Delta y = 0.
\end{aligned} \tag{C.10}$$

Then the velocity u has the form

$$u(i,j,l,t) = u^*(i,j,l,t) + \frac{S_\mathrm{p}}{a_\mathrm{p}}(p'(i,j,l,t) - p'(i-1,j,l,t)). \tag{C.11}$$

Therefore, substituting the velocity in x-, y- and z-direction into the discrete mass conservation equation (C.10), one has the discretiztion pressure correction term as

$$\begin{aligned} a_\mathrm{pp}p'(i,j,l,t) &= a_\mathrm{qw}q'(i-1,j,l,t) + a_\mathrm{qe}q'(i+1,j,l,t) + a_\mathrm{qs}q'(i,j-1,l,t) \\ &\quad + a_\mathrm{qn}q'(i,j_1,l,t) + a_\mathrm{qb}q'(i,j,l-1,t) + a_\mathrm{qt}q'(i,j,l+1,t) \\ &\quad + b_\mathrm{p}, \end{aligned} \tag{C.12}$$

where

$$\begin{aligned} a_\mathrm{pw} &= \frac{A_\mathrm{w}}{a_\mathrm{ue}}\Delta y \Delta z, \\ a_\mathrm{pe} &= \frac{A_\mathrm{e}}{a_\mathrm{uw}}\Delta y \Delta z, \\ a_\mathrm{ps} &= \frac{A_\mathrm{s}}{a_\mathrm{vs}}\Delta z \Delta x, \\ a_\mathrm{pn} &= \frac{A_\mathrm{n}}{a_\mathrm{vn}}\Delta z \Delta x, \\ a_\mathrm{pb} &= \frac{A_\mathrm{b}}{a_\mathrm{wb}}\Delta x \Delta y, \\ a_\mathrm{pt} &= \frac{A_\mathrm{t}}{a_\mathrm{wt}}\Delta x \Delta y, \\ a_\mathrm{pp} &= a_\mathrm{pw} + a_\mathrm{pe} + a_\mathrm{ps} + a_\mathrm{pn} + a_\mathrm{pb} + a_\mathrm{pt}, \\ b_\mathrm{p} &= (u^*(i,j,l,t) - u^*(i+1,j,l,t))\Delta y \Delta z, \\ &\quad +(v^*(i,j,l,t) - v^*(i,j+1,l,t))\Delta z \Delta x, \\ &\quad +(w^*(i,j,l,t) - w^*(i,j,l+1,t))\Delta x \Delta y, \end{aligned}$$

and define S_w, S_e, S_s, S_n, S_b and S_t

$$\begin{aligned} S_\mathrm{w} &= S_\mathrm{e} = \Delta y \Delta z, \\ S_\mathrm{s} &= S_\mathrm{n} = \Delta z \Delta x, \\ S_\mathrm{b} &= S_\mathrm{t} = \Delta x \Delta y. \end{aligned}$$

4. Correct u, v and w based on the correction term of the pressure. u is corrected according to (C.11), and v, w are corrected according to

$$u(i,j,l,t) = u^*(i,j,l,t) + \frac{S_\mathrm{w}}{a_\mathrm{w}}\Big(p'(i-1,j,l,t) - p'(i,j,l,t)\Big), \tag{C.13a}$$

$$v(i,j,l,t) = v^*(i,j,l,t) + \frac{S_\mathrm{s}}{a_\mathrm{s}}\Big(p'(i,j-1,l,t) - p'(i,j,l,t)\Big), \tag{C.13b}$$

$$w(i,j,l,t) = w^*(i,j,l,t) + \frac{S_\mathrm{b}}{a_\mathrm{b}}\Big(p'(i,j,l-1,t) - p'(i,j,l,t)\Big). \tag{C.13c}$$

5. Solve the temperature distribution problem based on equation (4.20), and calculate the water viscosity.

6. Treat the pressure field p shown in equation (C.12) as the guessed pressure field p^* and go back to step 2, then repeat the whole procedure, until a converged solution of p, u, v, w and T_f is achieved.

D Nomenclature

mathematical definitions

$\boldsymbol{n}$	outward normal of a surface
∇	gradient vector $\frac{\partial}{\partial x} + \frac{\partial}{\partial y} + \frac{\partial}{\partial z}$
∇^2	Laplace operator $\frac{\partial^2}{\partial x^2} + \frac{\partial^2}{\partial y^2} + \frac{\partial^2}{\partial z^2}$
t	time variable
t_c [s]	time constant of a system
t_s [s]	sample period
(x, y, z)	triangular rectangular coordinates system
$< \boldsymbol{a}, \boldsymbol{b} >$	inner product $\boldsymbol{a}^T \cdot \boldsymbol{b}$
$\mathscr{L}(\boldsymbol{X}, \boldsymbol{Y})$	bounded linear operator from Hilbert space $\boldsymbol{X}$ to Hilbert space $\boldsymbol{Y}$
$\lfloor x \rfloor$	floor i.e. $\lfloor x \rfloor$ is the largest integer smaller than or equal to $x \in \mathbb{R}$
$\frac{D}{Dt}(\cdot)$	material derivative, $\frac{D}{Dt}(\cdot) = \frac{d(\cdot)}{dt} + (\boldsymbol{v})\nabla(\cdot)$, with $\boldsymbol{v}$ is the velocity
$\frac{\partial(\cdot)}{\partial \boldsymbol{n}}$	gradient

descriptions of the die

N	number of cores
P_i [W]	power of core i
$V_{\mathrm{dd}i}$ [V]	supply voltage of core i
f_i [Hz]	working frequency of core i
Q [J]	heat/energy
q_i [W/m^3]	internal heat generation function of core i
S [m^2]	total closed surface of a volume
$\mathcal{S}_k$ [m^2]	k^{th} surface of the die
$\boldsymbol{n}_k$	normal vector on S_k
L_x, L_y, L_z [m]	size of the die in x-, y- and z-direction
Ω [m^3]	total die volume
$T(x, y, z, t)$ [K]	temperature distribution in the die
$T_0(x, y, z)$ [K]	initial temperature distribution function of the die
T_∞ [K]	ambient temperature

$q_i(x, y, z, t)$ [W/m^3]	internal heat generation function per unit time and per unit volume of core i
K [W/(m · K)]	thermal conductivity of the die
σ [J/(kg · K)]	specific heat capacitance of the die
ρ [kg/m^3]	density of the die
h [kg/(m^2 · K)]	convective heat transfer coefficient
h_{up}[kg/(m^2 · K)]	convective heat transfer coefficient of the primary heat escaping path
h_{down}[kg/(m^2 · K)]	convective heat transfer coefficient of the second heat escaping path

descriptions of the eigenvalue and eigenfunction system

$a,\ b,\ c$	index of eigenvalue in x-, y- and z-direction $\in \{0, 1, 2, ...\}$
$\phi_{xa},\ \phi_{yb},\ \phi_{zc}$	eigenfunctions in x-, y- and z-directions
$\phi_{abc}(x, y, z)$	eigenfunctions of die
$\lambda_{xa},\ \lambda_{yb},\ \lambda_{zc}$	eigenvalues in x-, y- and z-directions
λ_{abc}	eigenvalues of die

descriptions of the 1D PDEs and controller

Φ [W/m^2]	heat flux
$q_{\mathrm{ex},ij}$ [W/m^3]	energy generation function of core i due to the heat exchange with core j
$\boldsymbol{T}_a$ [K]	die average temperature defined by Definition 3.1
$\mathscr{T}_a$	state space, which is Hilbert space
$\boldsymbol{\xi}_1$	measurement output
Ξ_1	measurement output space, which is Hilbert space
$\boldsymbol{\xi}_2$	control output
Ξ_2	control output space, which is Hilbert space
$\mathscr{P}$	system input space, which is Hilbert space
$\boldsymbol{A}$	system operator
$\boldsymbol{B}$	input operator, $\mathscr{L}(\mathscr{P}, \mathscr{T}_{\mathrm{a}})$
$\mathscr{C}_1$	measurement output operator, $\mathscr{L}(\mathscr{T}_{\mathrm{a}}, \Xi_1)$
$\mathscr{C}_2$	regulated output operator, $\mathscr{L}(\mathscr{T}_{\mathrm{a}}, \Xi_2)$
$\boldsymbol{P}$	the input vector
$\boldsymbol{P}_{\mathrm{t}}$	target power consumption vector
$\boldsymbol{P}_{\mathrm{c}}$	the control input vector, and $\boldsymbol{P} = \boldsymbol{P}_{\mathrm{t}} + \boldsymbol{P}_{\mathrm{c}}$

descriptions of the micro-channel

$\boldsymbol{V}_{\mathrm{s}}$ [m/s]	liquid velocity
(u, v, w) [m/s]	the liquid velocity in x-, y- and z-direction
u_{in} [m/s]	channel inlet liquid velocity
p [N/m^2]	average pressure
ι [N/m^2]	body force
τ_{ii} [N/m^2]	normal stress
τ_{ij} [N/m^2]	shear stresses
ρ_{f} [kg/m^3]	density of the liquid
h_{f} [kg/m^3]	convective heat transfer coefficient between the liquid and its surrounding
K_{f} $[W/(M \cdot K)]$	thermal conductivity of the liquid
σ_{f} $[J/(kg \cdot K)]$	specific heat capacity of the liquid
μ [Pa · s]	water viscosity
$\boldsymbol{T}_{\mathrm{f}}$ [K]	temperature vector of the liquid
T_{in} [K]	channel input liquid temperature
W_{ch} [m]	width of the channel
H_{ch} [m]	height of the channel
D_e [m]	channel hydraulic diameter
L_{f} [m]	length of the fluid entrance region
L_{t} [m]	length of the thermal entrance region
P_{f} [W]	heat flow
m_{f} [kg/s]	mass flow
H [J/kg]	specific enthalpy
$\mathcal{H}$ [J]	enthalpy
P_{net} [W]	work per time unit/ power
g [m/s^2]	gravitational acceleration
ΔL_i [m]	length of the channel block in x-direction
Re	Reynolds number
Pr	Prandtl number
Nu	Nusselt number

descriptions of the R-C model

R [K/W]	thermal resistance
C [J/K]	thermal capacitance
$\boldsymbol{T}_{\mathrm{d}}$ [K]	temperature vector of the dies

descriptions of the 3D stacked package model

$\boldsymbol{T}(t)$	system state vector $\boldsymbol{T}(t) = \left[\boldsymbol{T}_{\mathrm{d}}^T(t), \boldsymbol{T}_{\mathrm{f}}^T(t)\right]^T$
$\boldsymbol{d}(t)$	unknown input vector

$\boldsymbol{A}$	system matrix
$\boldsymbol{B}_{(\cdot)}$	input matrix
$\boldsymbol{C}_{(\cdot)}$	output matrix
$\boldsymbol{L}_{\delta(k)}$	observer gain matrix
$\delta(k)$	switching law
$(\cdot)_{\mathrm{d},\cdot}$	matrix with a index 'd' is for discrete model
$\widehat{(\cdot)}$	describe relative to prediction
$\check{(\cdot)}$	describe relative to the new system combining system and observer error dynamic

abbreviation

MCP	multi-core processor
1D	one dimensional
2D	two dimensional
3D	three dimensional
I/O	input/output
CMOS	Complementary metal-oxide-semiconductor
PMOS	P-type mental-oxide-semiconductor
NMOS	N-type mental-oxide-semiconductor
PDE	partial differential equation
ODE	ordinary differential equation
R-C	thermal resistance and capacitance
TSV	through-Silicon via
MCLCS	micro-channel liquid cooling system
MPC	model prediction control
SIMPLE	Semi-implicit method for pressure linked equations

Bibliography

[AAF98] S. F. Al-Sarawi, D. Abbott, and P. D. Franzon. A review of 3-D packaging technology. *IEEE Transactions on Components, Packaging, and Manufacturing Technology - Part B*, 21:2–14, 1998.

[AB10] M. L. Abell and J. P. Braselton. *Introductory Differential Equations: With Boundary Value Problems, Third Edition.* Elsevier, 2010.

[AFP04] A. Abdollahi, F. Fallah, and M. Pedram. Leakage current reduction in CMOS VLSI circuits by input vector control. *IEEE Transactions on Very Large Scale Integration (VLSI) Systems*, 12:140–154, 2004.

[ASP+09] J. L. Ayala, A. Sridhar, V. Pangracious, D. Atieza, and Y. Leblebici. Through silicon via-based grid for thermal control in 3D chips. *4th International Conference on Nano-Networks*, pages 90–98, 2009.

[Bak10] R. J. Baker. *COMS - Circuit Design, Layout, and Simulation.* John Wiley & Sons, 2010.

[Bat00] G. K. Batchelor. *An Introduction to Fluid Dynamics.* Cambridge University Press, 2000.

[BEFB94] S. Boyd, L. El Ghaoui, E. Feron, and V. Balakrishnan. *Linear Matrix Inequalities in System and Control Theory.* Society for Industrial and Applied Mathematics (SIAM), 1994.

[BILD90] T. L. Bergman, F. P. Incropera, A. S. Lavine, and D. P. DeWitt. *Fundamentals of Heat and Mass Transfer (3rd ed.).* John Wiley & Sons, 1990.

[CAA+09] A. K. Cockum, J. L. Ayala, D. Atienya, T. S. Rosing, and Y. Leblebichi. Dynamic thermal management in 3D multicore architectures. *Design, Automation and Test in Europe Conference and Exhibition*, pages 1410–1415, 2009.

[CAAR09] A. K. Coskun, J. L. Ayala, D. Atienza, and T. S. Rosing. Modeling and dynamic management of 3D multicore systems with liquid cooling. *17th IFIP International Conference on very Large Scale Integration (VLSI-SoC)*, pages 60–65, 2009.

[CAR+10] A. K. Coskun, D. Atienza, T. S. Rosing, T. Brunschwiler, and B. Michel. Energy-efficient variable-flow liquid cooling in 3D stacked architectures.

Design, Automation and Test in Europe Conference and Exhibition, pages 111–116, 2010.

[CB95] A. P. Chandrakasan and R. W. Brodersen. Minimizing power consumption in digital CMOS circuits. *Proceedings of the IEEE*, 83:498–523, 1995.

[CB07] E. F. Camacho and C. Bordons. *Model Predictive Control*. Springer, 2007.

[CCC12] H.-C. Cheng, I.-C. Chung, and W.-H. Chen. Thermal chip placement in MCMs using a novel hybrid optimization algorithm. *IEEE Transactions on Components, Packaging and Manufacturing Technology*, 2:764–774, 2012.

[CPLK12] H. J. Choi, Y. J. Park, H.-H. Lee, and C. H. Kim. Adaptive dynamic frequency scaling for thermal-aware 3D multi-core processors. *Lecture Notes in Computer Science*, 7336:602–612, 2012.

[CRAI13] D. Cuesta, J. L. Risco-Martin, J. L. Ayala, and J. Ignacio Hidalgo. 3D thermal-aware floor planner using a MOEA approximation. *VLSI Integration Journal*, 46:10–11, 2013.

[CRT98] Y.-K. Cheng, P. Raha, and C.-C. Teng. ILLIADS-T: An electrothermal timing simulator for temperature-sensitive reliability diagnosis of CMOS VLSI chips. *IEEE Transactions on Computer-Aided Design of Integrated Circuits and Systems*, 6:668–681, 1998.

[CZ95] R. F. Curtain and H. J. Zwart. *An Introduction to Infinite-Dimensional Linear Systems Theory*. Springer Verlag, 1995.

[DMNH10] A. Doostaregan, A. H. Moaiyeri, K. Navi, and O. Hashemipour. On the design of new low-power CMOS standard ternary logic gates. *15th CSI International Symposium on Computer Architecture and Digital Systems*, pages 115–120, 2010.

[Dun44] W. J. Duncan. Some devices for the solution of large sets of simultaneous linear equations. *Philosophical Magazine Series 7*, 35:660–670, 1944.

[DW04] J. Dabrowski and E. R. Weber. *Predictive Simulation of Semiconductor Processing: Status and Challenges*. Springer Verlag, 2004.

[Fey60] R. Feynman. There's plenty of room at the bottom. *Engineering and Science*, 23:22–36, 1960.

[FKLK12] Y. Fu, N. Kottenstette, C. Lu, and X. D. Koutsoukos. Feedback thermal control of real-time systems on multicore processors. *Proceedings of the tenth ACM international conference on Embedded software*, pages 113–122, 2012.

[Fou09] J. B. J. Fourier. *The Analytical Theory of Heat*. Cambridge University Press, 2009.

[Fre09] Freescale. MC13783 data sheet. *Freescale Semiconductor*, 2009.

[Gad99] M. Gad-el-Hak. The fluid mechanics of microdevices - the Freeman scholar lecture. *Journal of fluids engineering*, 121:5–33, 1999.

[Geb71] B. Gebhart. *Heat Transfer*. McGraw-Hill Book Company, 1971.

[Gee05] D. Geer. Chip makers turn to multicore processors. *Computer*, 38:11–13, 2005.

[GK06] P. Gepner and M. F. Kowalik. Multi-core processors: new way to achieve high system performance. *Proceedings of the International Symposium on Parallel Computing in Electrical Engineering*, pages 9–13, 2006.

[Har06] M. Hartman. Powerwise adaptive voltage scaling minimizes energy consumption. *Texas Instruments Incorporated Technical Report*, 2006.

[HKG99] T. M. Harms, M. J. Kazmierczak, and F. M. Gerner. Developing convective heat transfer in deep rectangular microchannels. *International Journal of heat and Fluid Flow*, 20:149–157, 1999.

[HL09] P.-Y. Huang and Y.-M. Li. Full-chip thermal analysis for the early design stage via generalized integral transforms. *IEEE Transactions on Very Large Scale Integration Systems*, 17:613–626, 2009.

[Hol86] J. P. Holman. *Heat Transfer, Sixth Edition*. McGraw-Hill Book Company, 1986.

[Hol12] E. Holzbecher. *Environmental modeling: using MATLAB*. Springer, 2012.

[HRW11] D. Halliday, R. Resnick, and J. Walker. *Fundamentals of Physics Extended, 9th Edition*. Wiley, 2011.

[Hwa06] E. O. Hwang. *Digital Logic and Microprocessor Design with VHDL*. Thomson, 2006.

[Int10a] Intel. Intel® core™ i7-600, i5-500, i5-400 and i3-300 mobile processor series. *http://download.intel.com/design/processor/datashts/322812.pdf*, 2010.

[Int10b] Intel. Intel® core™ i7-900 desktop processor extreme edition series and intel® core™ i7-900 desktop processor series. *http://www.intel.com/content/dam/www/public/us/en/documents/datasheets/core-i7-900-ee-and-desktop-processor-series-datasheet-vol-1.pdf*, 2010.

[JM09a] R. Jayaseelan and T. Mitra. Dynamic thermal management via architectural adaptation. *46th ACM/IEEE Design Automation Conference*, pages 484–489, 2009.

[JM09b] R. Jayaseelan and T. Mitra. A hybrid local-global approach for multi-core thermal management. *IEEE/ACM International Conference on Computer-Aided Design*, pages 314–320, 2009.

[Kah99] J. Kahle. Power4: A dual-CPU processor chip. *Microprocessor Forum*, 1999.

[KB95] A. D. Kraus and A. Bar-Cohen. *Design and analysis of heat sinks.* John Wiley & Sons, 1995.

[KC05] W. M. Kays and M. Crawford. *Convective heat and mass transfer, 4th Edition.* McGraw-Hill, 2005.

[KIJG05] J.-M. Koo, S. Im, L. Jiang, and K. E. Goodson. Integrated microchannel cooling for three-dimensional electronic circuit architectures. *Journal of Heat Transfer*, 127:49–58, 2005.

[Kin12] C. R. King Jr. *Thermal Management of Three-Dimensional Integrated Circuits Using Inter-Layer Liquid.* PhD thesis, Georgia Institute of Technology, 2012.

[KMS13] F. Kreith, R. M. Manglik, and Bohn M. S. *Principles of Heat Transfer, SI Edition.* CL Engineering,7 edition., 2013.

[Kre00] F. Kreith. *The CRC Handbook of Thermal Engineering.* Springer, 2000.

[KZYW09] C. Kuang, W. Zhao, F. Yang, and G. Wang. Measuring flow velocity distribution in microchannels using molecular tracers. *Microfluid Nanofluid*, 7:509–517, 2009.

[Lai10] Laing. Laing 12 volt DC pumps datasheets. *Laing GmbH*, 2010.

[Löf04] J. Löfberg. Yalmip: A toolbox for modeling and optimization in matlab. *IEEE International Symposium on Computer Aided Control Systems Design*, pages 284–289, 2004.

[LFQ12] G. Liu, M. Fan, and G. Quan. Neighbor-aware dynamic thermal management for multi-core platform. *Design, Automation & Test in Europe Conference & Exhibition*, 2012.

[LK99] Y. I. Lee and B. Kouvaritakis. Constrained receding horizon predictive control for systems with disturbances. *International Journal of Control*, 72:1027–1032, 1999.

[LK01] Y. I. Lee and B. Kouvaritakis. Receding horizon output feedback control for linear systems with input saturation. *IEE Proceedings Control Theory & Applications*, 148:109–115, 2001.

[Lov10] R. Love. *Linux Kernel Development.* Addison-Wesley, 2010.

[LY09] J. H. Lau and T. G. Yue. Thermal management of 3D IC integration with TSV (through silicon via). *59th Electronic Components and Technology Conference*, pages 635–640, 2009.

[MDVPO90] P. Macken, M. Degrauwe, M. Van Paemel, and H. Oguey. A voltage reduction technique for digital systems. *IEEE International Solid-State Circuits Conference*, pages 238–239, 1990.

[MMA+08] S. Murali, A. Mutapcic, A. Atienza, R. Gupta, S. Boyd, L. Benini, and G. De Micheli. Temperature control of high-performance multi-core platforms using convex optimization. *Design, Automation and Test in Europe*, pages 110–115, 2008.

[MÖ94] M. D. Mikhailov and M. N. Özisik. *Unified Analysis and Solutions of Heat and Mass Diffusion.* Dover Publications, Inc., 1994.

[Mor13] F. A. Morrison. *Introduction to Fluid Mechans.* Cambridge University Press, 2013.

[Mot09] M. Motoyoshi. Through-silicon via. *Proceedings of the IEEE*, 97:43–48, 2009.

[MSS+05] P. Michaud, Y. Sazeides, A. Seznec, T. Constantinou, and D. Fetis. An analytical model of temperature in microprocessors. *Research Report, RR5744, IRSIA*, 2005.

[MYL09] H. Mizunuma, C.-L. Yang, and Y.-C. Lu. Thermal modeling for 3D-ICs with integrated microchannel cooling. *IEEE/ACM International Conference on Computer-Aided Design*, pages 256–264, 2009.

[Ölç64] N. Y. Ölçer. On the theory of conductive heat transfer in finite regions. *International Journal of Heat and Mass Transfer*, 7:307–314, 1964.

[ORCP93] A. Ortega, S. L. Ramanathan, J. D. Chicci, and J. L. Prince. Thermal wake models for forced air cooling of electronic components. *Ninth Annual IEEE Semiconductor Thermal Measurement and Management Symposium*, pages 63–74, 1993.

[Pat80] S. V. Patankar. *Numerical Heat Transfer and Fluid Flow.* McGraw Hill Book Company, 1980.

[PBB+05] D. Pham, E. Behnen, M. Bolliger, H. P. Hofstee, C. Johns, J. Kahle, A. Kameyama, J. Keaty, B. Le, Y. Masubuchi, S. Posluszny, M. Riley, M. Suzuoki, M. Wang, J. Warnock, S. Weitzel, D. Wendel, and K. Yazawa. The design methodology and implementation of a first generation cell processor: a multi-core SoC. *IEEE Custom Integrated Circuits Conference*, 2005.

[Pol01] A. D. Polyanin. *Handbook of Linear Partial Differential Equations for*

Engineers and Scientists. Champman & Hall/CRC, 2001.

[QM02] W. Qu and I. Mudawar. Analysis of three dimensional heat transfer in micro-channel heat sinks. *International Journal of Heat and Mass Transfer*, 45:3973–3985, 2002.

[QM03] W. Qu and I. Mudawar. Thermal design methodology for high-heat-flux single-phase and two-phase micro-channel heat sinks. *IEEE Transactions on Components and Packaging Technologies*, 26:589–609, 2003.

[Rat06] J. Rattner. Why multi-core? *Intel Developer Forum*, 2006.

[SAS02] K. Skadron, T. Abdelzaher, and M. R. Stan. Control-theoretic techniques and thermal-RC modeling for accurate and localized dynamic thermal management. *Proceedings of Eighth International Symposium on High-Performance Computer Architecture*, pages 17–28, 2002.

[SB87] R. K. Shah and M. S. Bhatti. Laminar convective heat transfer in ducts. In S. Kakac, R.K. Shan, and W. Wung, editors, *Handbook of Single-Phase Convective Heat Transfer.* Willy, 1987.

[SB01] S. K. Som and G. Biswas. *Introduction to Fluid Mechanics & Fluid Machines: 2nd.* McGraw-Hill, 2001.

[Sch10] M. T. Schobeiri. *Fluid Mechanics for Engineers: A Graduate Textbook.* Springer, 2010.

[SL78] R. K. Shah and A. L. London. *Advances in Heat Transfer Supplement I: Laminar Flow Forced Convection in Ducts.* Academic Press, 1978.

[SLS09] M. R. Spiegel, S. Lipcshutz, and D. Spellman. *Schaum's Outline of Vector Analysis.* McGraw Hill, 2009.

[SSH+03a] K. Skadron, M. R. Stan, W. Huang, S. Velusamy, K. Sankaranarayanan, and Tarjan D. Temperature-aware microarchitecture. *Proceedings of 30th Annual International Symposium on Computer Architecture*, pages 2–13, 2003.

[SSH+03b] K. Skadron, M. R. Stan, W. Huang, S. Velusamy, K. Sankaranarayanan, and D. Tarjan. Temperature-aware microarchitecture: extended discussion and results. Technical report, University of Virginia, Department of Computer Science, CS-2003-08, 2003.

[Stu99] J. F. Sturm. Using sedumi 1.02, a matlab toolbox for optimization over symmetric cones. *Optimization Methods and Software*, 11-12:545–581, 1999.

[SVS06] O. Semenov, A. Vassighi, and M. Sachdev. Impact of self-heating effect on long-term reliability and performance degradation in CMOS circuits.

IEEE Transactions on Device and Materials Reliability, 6:17–27, 2006.

[Tay52] A. E. Taylor. L'Hospital's rule. *The American Mathematical Monthly*, 59:20–24, 1952.

[Til13] Tilera. TILE-Gx8072 datasheet. http://www.tilera.com/sites/default/files/productbriefs/TILE-Gx8072£_£PB041-02.pdf, 2013.

[VS06] A. Vassighi and M. Sachdev. *Thermal and Power Management of Integrated Circuits*. Springer, 2006.

[VWWL00] R. Viswanath, V. Wakharkar, A. Watwe, and V. Lebonheur. Thermal performance challenges from silicon to systems. *Intel Technology Journal*, Q3, 2000.

[WB12] B. Wojciechowski and M. A. Bawiec. Practical dynamic thermal management of multi-core microprocessors. *18th International Workshop on Thermal Investigations of ICs and Systems*, pages 1–4, 2012.

[WJ02] X. Wei and Y. Joshi. Optimization study of stacked micro-channel heat sink for micro-electronic cooling. *The Eighth Intersociety Conference on Thermal and Thermomechanical Phenomena in Electronic Systems*, pages 55 – 61, 2002.

[WLC03] T.-Y. Wang, Y.-M. Lee, and C. C.-P. Chen. 3D thermal-ADI - an efficient chip level transient thermal simulator. *Proceeding of International Symposium on Physical Design*, pages 10–17, 2003.

[ZAD09] F. Zanini, D. Atienza, and D. De Micheli. A control theory approach for thermal balancing of MPSoC. *Asia and South Pacific Design Automation Conference*, pages 37–42, 2009.

[ZAD13] F. Zanini, D. Atienza, and G. De Micheli. A combined sensor placement and convex optimization approach for thermal management in 3D-MPSoC with liquid cooling. *VLSI Integration Journal*, 46:33–43, 2013.

[ZDG06] K. Zhou, J. C. Doyle, and K. Glover. *Robust and Optimal Control*. Prentice-Hall, 2006.

[ZTH09] C. Zhang, G. Tang, and S. Han. Approximate design of optimal tracking controller for systems with delayed state and control. *IEEE International Conference on Control and Automation*, pages 1168–1172, 2009.

[ZXD$^+$08] X. Zhou, Y. Xu, Y. Du, Y. Zhang, and J. Yang. Thermal management for 3D processors via task scheduling. *37th International Conference on Parallel Processing*, pages 115–122, 2008.

Jianfei Wang

Education

08/2008-01/2014	Studying for Doctor of Engineering, Chair of Control System Kaiserslautern University of Technology (TU-KL), Germany
09/2005-04/2008	Master of Science, College of Automation Engineering Nanjing University of Aeronautics and Astronautics (NUAA), China
09/2001-08/2005	Bachelor of Science, College of Electrical Engineering Hohai University (HHU), China

Practical trainings and work experiences

01/2009-10/2013	Furuta rotary pendulum experiment system, TU-KL, Germany
09/2005-02/2006	Development of automatic production line and warehouse experiment system, NUAA, China

Publication

[1] Jianfei Wang, Steven Liu, Modeling and control for thermal balancing of multi-core processors, *Journal of The Franklin Institute*, vol.350, no.7, pp. 1836-1847, 2013.

In der Reihe „*Forschungsberichte aus dem Lehrstuhl für Regelungssysteme*“, herausgegeben von Steven Liu, sind bisher erschienen:

1	Daniel Zirkel	Flachheitsbasierter Entwurf von Mehrgrößenregelungen am Beispiel eines Brennstoffzellensystems ISBN 978-3-8325-2549-1, 2010, 159 S. 35.00 €
2	Martin Pieschel	Frequenzselektive Aktivfilterung von Stromoberschwingungen mit einer erweiterten modellbasierten Prädiktivregelung ISBN 978-3-8325-2765-5, 2010, 160 S. 35.00 €
3	Philipp Münch	Konzeption und Entwurf integrierter Regelungen für Modulare Multilevel Umrichter ISBN 978-3-8325-2903-1, 2011, 183 S. 44.00 €
4	Jens Kroneis	Model-based trajectory tracking control of a planar parallel robot with redundancies ISBN 978-3-8325-2919-2, 2011, 279 S. 39.50 €
5	Daniel Görges	Optimal Control of Switched Systems with Application to Networked Embedded Control Systems ISBN 978-3-8325-3096-9, 2012, 201 S. 36.50 €
6	Christoph Prothmann	Ein Beitrag zur Schädigungsmodellierung von Komponenten im Nutzfahrzeug zur proaktiven Wartung ISBN 978-3-8325-3212-3, 2012, 118 S. 33.50 €
7	Guido Flohr	A contribution to model-based fault diagnosis of electro-pneumatic shift actuators in commercial vehicles ISBN 978-3-8325-3338-0, 2013, 139 S. 34.00 €